T0331086

Self-Healing
Cementitious Materials

Emerging Materials and Technologies
Series Editor
Boris I. Kharissov

Recycling of Plastics, Metals, and Their Composites
R.A. Ilyas, S.M. Sapuan, and Emin Bayraktar

Viral and Antiviral Nanomaterials
Synthesis, Properties, Characterization, and Application
Devarajan Thangadurai, Saher Islam, Charles Oluwaseun Adetunji

Drug Delivery using Nanomaterials
Yasser Shahzad, Syed A.A. Rizvi, Abid Mehmood Yousaf and Talib Hussain

Nanomaterials for Environmental Applications
Mohamed Abou El-Fetouh Barakat and Rajeev Kumar

Nanotechnology for Smart Concrete
Ghasan Fahim Huseien, Nur Hafizah A. Khalid, and Jahangir Mirza

Nanomaterials in the Battle Against Pathogens and Disease Vectors
Kaushik Pal and Tean Zaheer

MXene-Based Photocatalysts: Fabrication and Applications
Zuzeng Qin, Tongming Su, and Hongbing Ji

Advanced Electrochemical Materials in Energy Conversion and Storage
Junbo Hou

Emerging Technologies for Textile Coloration
Mohd Yusuf and Shahid Mohammad

Emerging Pollutant Treatment in Wastewater
S.K. Nataraj

Heterogeneous Catalysis in Organic Transformations
Varun Rawat, Anirban Das, Chandra Mohan Srivastava

2D Monoelemental Materials (Xenes) and Related Technologies: Beyond Graphene
Zongyu Huang, Xiang Qi, Jianxin Zhong

Atomic Force Microscopy for Energy Research
Cai Shen

Self-Healing Cementitious Materials: Technologies, Evaluation Methods, and Applications
Ghasan Fahim Huseien, Iman Faridmehr, Mohammad Hajmohammadian Baghban

For more information about this series, please visit:
www.routledge.com/Emerging-Materials-and-Technologies/book-series/CRCEMT

Self-Healing Cementitious Materials

Technologies, Evaluation Methods, and Applications

Ghasan Fahim Huseien, Iman Faridmehr,
Mohammad Hajmohammadian Baghban

CRC Press
Taylor & Francis Group
Boca Raton   London

CRC Press is an imprint of the
Taylor & Francis Group, an **informa** business

First edition published 2022
by CRC Press
6000 Broken Sound Parkway NW, Suite 300, Boca Raton, FL 33487–2742

and by CRC Press
2 Park Square, Milton Park, Abingdon, Oxon, OX14 4RN

© 2022 Taylor & Francis Group, LLC

CRC Press is an imprint of Taylor & Francis Group, LLC

ISBN: 978-1-032-05038-6 (hbk)
ISBN: 978-1-032-05039-3 (pbk)
ISBN: 978-1-003-19576-4 (ebk)

DOI: 10.1201/9781003195764

Typeset in Times
by Apex CoVantage, LLC

Contents

Preface.. ix
Biographies xi

Chapter 1 Cementitious Construction Materials Sustainability:
An Introduction 1

    1.1   Introduction 1
    1.2   Cement-Generated Environmental Problems 2
    1.3   Energy Problems in Cement Industries 2
    1.4   Concrete Performance in Aggressive Environments 3
    1.5   Crack Problems in Concrete.. 4
    1.6   Sustainability of Smart Concrete... 5
    1.7   Effect of Addition Self-Healing Agents on
Concrete Properties 5
    1.8   Life Cycle Analysis of Self-Healing Concrete 6
    1.9   Summary.. 6

Chapter 2 Self-Healing Technology: Fundamental and Design Strategies 11

    2.1   Introduction 11
    2.2   Sustainability of Self-Healing Concrete.................................. 12
    2.3   Mechanisms of Self-Healing in Cementitious Materials......... 14
    2.4   Design Strategies 14
            2.4.1   Release of Healing Agent... 15
            2.4.2   Reversible Cross-Links 16
            2.4.3   Miscellaneous Technologies....................................... 17
    2.5   Summary.. 17

Chapter 3 Self-Healing Cementitious Materials 21

    3.1   Introduction 21
    3.2   Self-Healing Processes in Concrete... 21
    3.3   Expansive Agents and Mineral Admixtures 24
    3.4   Hollow Fibers 24
    3.5   Bacteria as Self-Healing Agent 27
    3.6   Microencapsulation 27
    3.7   Shape Memory Materials as Self-Healer.................................. 29
    3.8   Coating 30
    3.9   Engineered Cementitious Composite 31
    3.10   Nanomaterials-Based Self-Healing Concrete.......................... 33
    3.11   Self-Healing in Fiber-Reinforced Concrete............................. 33
    3.12   Summary.. 34

Chapter 4   Self-Healing Measurement Methods.. 41

           4.1   Concrete Durability 41
           4.2   Self-Healing Techniques.. 42
           4.3   Self-Healing Measurement Methods.. 45
           4.4   Efficacy of Autonomous Healing Techniques 46
           4.5   Structure Tests for Evaluating Self-Healing Efficiency.......... 46
                   4.5.1   Macrostructure Tests 47
                   4.5.2   Microstructure Tests 48
                   4.5.3   Nanostructure Tests 48
           4.6   Comparison of Assessment Tests 49
           4.7   Summary.. 50

Chapter 5   Polymers-Based Self-Healing Cementitious Materials..................... 55

           5.1   Introduction 55
           5.2   Mix Design and Specimens' Preparation 57
           5.3   Fresh Properties 60
           5.4   Strength Properties 61
           5.5   Self-Healing Performance 68
           5.6   Developing an ANN to Estimate Degree of Damage
                  and Healing Efficiency 74
                   5.6.1   Firefly Optimization Algorithm (FOA) 76
                   5.6.2   Generation of Training and Testing Data Sets........... 77
                   5.6.3   Results 79
           5.7   Summary.. 82

Chapter 6   Bacteria-Based Self-Healing Concrete... 87

           6.1   Introduction 87
           6.2   Self-Healing Approach 88
           6.3   Effect of Bacteria.. 90
           6.4   Summary.. 93

Chapter 7   Nanomaterials-Based Self-Healing Cementitious Materials............. 97

           7.1   Introduction 97
           7.2   Significance of Nanomaterials as Self-Healing...................... 97
           7.3   Production of Nanomaterials.. 97
           7.4   Nanomaterials-Based Concretes... 99
           7.5   Production of Nanoconcrete 99
           7.6   Nanomaterials-Based Self-Healing Concrete......................... 100
           7.7   Nanosilica... 101
           7.8   Nanoalumina.. 102
           7.9   Carbon Nanotube... 102
           7.10   Polycarboxylates... 103
           7.11   Titanium Oxide.. 104

7.12   Nanokaolin 104
7.13   Nanoclay 105
7.14   Nanoiron 106
7.15   Nanosilver 106
7.16   Summary 107

Chapter 8   Factors Affecting Concrete Self-Healing Performance 113

8.1   Impact of Crystalline Admixtures on Self-Healing 113
8.2   Polymers as Self-Healing Material 116
8.3   Fibers as Self-Healing Material 118
8.4   Self-Healing Performance 120
8.5   Summary 121

Chapter 9   Encapsulation Technology-Based Self-Healing Cementitious
Materials 123

9.1   Introduction 123
9.2   Bio-Based Healing Agents: Approaches
and Mechanisms 124
9.3   Incorporation of Encapsulated Healing Agents 126
9.4   Evaluation of Bio-Based Self-Healing Systems 128
9.5   Summary 130

Chapter 10   Applications, Future Directions, and Opportunities of Self-
Healing Concrete 133

10.1   Introduction 133
10.2   Sustainability of Smart Materials-Based Self-Healing
Concrete and Nanotechnology 134
10.3   Merits and Demerits of Nanomaterials for Self-Healing
Concrete 134
10.4   Economy of Nanomaterials-Based Self-Healing
Concretes 135
10.5   Environmental Suitability and Safety Features
of Nanomaterials-Based Concretes 136
10.6   Summary 137

Index 139

Preface

Climate change is anticipated to have a major impact on concrete structures through increasing rates of deterioration, as well the impacts of extreme weather events. The deterioration that occurs from the very beginning of the service not only reduces the lifespan of the concrete, but also demands more cement to maintain the durability. Meanwhile, the repair process of the damaged parts is highly labor-intensive and expensive. Thus, the self-healing of such damages is essential for environmental safety and energy cost savings.

Nowadays, cementitious self-healing materials are one of the most significant scientific and industrial breakthroughs of the twenty-first century. Cementitious self-healing materials offer great advantages toward concrete sustainability in construction fields such as energy storing, high performance, corrosion resistance, environmental remediation, and long life concrete applications. Self-healing properties can improve in concrete using different techniques including encapsulation of polymers, inclusion of fibers, and secondary hydration of unhydrated cement.

Chapter 1 deals with the sustainability of cementitious construction materials. The fundamental and design strategies of self-healing technology are widely discussed in Chapter 2. In Chapter 3, we focus on self-healing cementitious materials such as expensive agents, polymer, fiber, mineral admixtures, bacteria, and nanomaterials. The measurement methods which was adopted to evaluate the self-healing performance and efficiency are presented in Chapter 4. The applications of polymer, bacteria, nanomaterials, and encapsulated materials as cementitious self-healing agents are widely discussed in Chapter 5, Chapter 6, Chapter 7, and Chapter 9, respectively. Topics regarding factors affecting self-healing performance, sustainability, and benefits of self-healing technology are presented in Chapter 8 and Chapter 10.

With continued development and seamless integration of "self-healing" materials into modern-day applications, this is set to open up a new era of enhanced consumer experience, infrastructure maintenance, and environmental management, as well as countless unprecedented and unique applications resembling those in science fiction.

Ghasan Fahim Huseien
Iman Faridmehr
Mohammad Hajmohammadian Baghban

Biographies

Dr. Ghasan Fahim Huseien
Ghasan is a research associate at the Department of the Built Environment, School of Design and Environment, National University of Singapore. He has more than five years of applied research and development experience, as well as up to ten years' experience in manufacturing smart materials for sustainable building and smart cities. He has expertise in advanced sustainable construction materials covering civil engineering, environmental sciences, and engineering. He has authored and co-authored more than 75 publications and technical reports, four books, and 17 book chapters, and he has participated in more than 30 national and international conferences/workshops. His past experience in projects include application of nanotechnology in construction and building materials, self-healing technology, and geopolymers as sustainable and eco-friendly repair materials in the construction industry.

Dr. Iman Faridmehr
Iman is a senior research assistant at the South Ural State University, Russian Federation. His research agenda focuses on the cost-effective design of structures in compliance with sustainable design strategies, using low embodied energy materials and novel construction techniques. He has been involved in many collaborative research projects with researchers from different universities, including the Norwegian University of Science and Technology (NTNU), the National University of Singapore (NUS), and Universiti Teknologi Malaysia (UTM). Iman is (co-)author of 19 journal papers indexed by the Web of Science Core Collection. He provides service as a reviewer board for scientific and scholarly journals, including *Archives of Civil and Mechanical Engineering*, *Engineering Structures*, *Advances in Structural Engineering*, and *Construction and Building Materials*.

Assoc. Prof. Dr. Mohammad Hajmohammadian Baghban
Mohammad is an associate professor in building and construction technology at the Norwegian University of Science and Technology. He is a well-experienced researcher on developing sustainable and multi-functional building materials and components. He is also skilled in modeling and optimization methods, including the application of evolutionary optimization methods and machine learning in civil engineering. Presently, he is the director of the Building Material Laboratory at his

department at NTNU, and an active member of Norwegian Concrete Association (NCA), Nordic Concrete Federation (NCF), and the European Construction, Built Environment and Energy-Efficient Building Technology Platform (ECTP). He is (co-)author of more than 30 peer-reviewed articles in his field and has served as a review board member in relevant conferences and journals from different publishers, including Springer, MDPI, and Elsevier.

1 Cementitious Construction Materials Sustainability
An Introduction

1.1 INTRODUCTION

Cement industries that produce the main constituent of concrete remain the foremost concern to the world in terms of contribution for climate change and hindrance of sustainability. Apart from the United States, the fast-developing nations such as China, Indonesia, India, and Turkey are also causing major environmental pollution [1–2]. On [9], it was reported there are more than 4 billion metric tons of ordinary Portland cement (OPC) that is produced annually from numerous cement industries worldwide as a basic constituent of concrete are primarily responsible for the major CO_2 footprint [3–6]. Identification of satisfactory and realistic alternatives is a challenging task. In the civil engineering sector, OPC is broadly used as an efficient binder in concrete and other construction materials. Meanwhile, OPC manufacturing is commonly accepted as the main provider of emitted greenhouse gases in the atmosphere [7–10]. The International Energy Agency (IEA) reports suggest that it amounts to 6–7% of total CO_2 emissions [10–13]. By the year 2050, the worldwide demand of OPC is expected to boost nearly 200% [10]. Mitigation of CO_2 emissions from OPC-related activities requires new types of sustainable, smart, and environmental affable self-healing materials [14–16].

It is known that cement materials have very low resistance to aggressive environments, and that is one of the most severe problems affecting the durability and service life of concrete structures in the natural climate. The effect of aggressive environment can manifest in the form of expansion and cracking of concrete. Sometimes the cracking of concrete may cause serious structural problems [17–18]. In this sense, the continuous increase in the durability requirements of the concrete structures led to the development of smart concretes (using self-healing technology) where excellent durability is mandatory [19].

Generally, self-healing is beneficial for the materials' durability. Particularly, it is advantageous when human interference is difficult, such as in construction purposes in the midst of harsh physical and chemical environments. Self-healing is also required to protect material characteristics, especially in kinetic and thermodynamic conditions that support large defects density like nanostructures. Nanomaterials invariably reveal excellent functional attributes. Compared to ordinary materials, nanomaterials degrade faster due to the presence of numerous interfacial atoms.

DOI: 10.1201/9781003195764-1

Many functional nanostructures can be combined to fabricate diverse nanosystems, wherein some components can also be incorporated to offer self-healing actions. In fact, this strategy is simplistic compared to the design of a more robust nanosystem [20]. Of late, due to the rapid advancement of nanoscience and nanotechnology, the design and fabrication of self-healing materials has taken new frontiers, wherein materials with particle size below 500 nm are termed as nanomaterials. Self-healing materials can recover from the damages autonomously. In many circumstances, the self-healing action can also be prompted via temperature as external stimuli; systems with this capacity are called non-autonomic self-healing materials [21].

Looking at the future prospects self-healing concrete, this book revisits the ongoing research activities on nanomaterials-based self-healing concrete beneficial for future sustainable developments. The present chapter is organized in three main sections. Section 1.2 and Section 1.3 discuss the energy and environmental problems of the cement and concrete industry. Section 1.4 and Section 1.5 highlight the cementitious construction materials performance in aggressive environments and the effect of cracks on concrete structures' lifespans. Section 1.6, Section 1.7, and Section 1.8 deal with the importance of self-healing concrete toward more sustainable and smarter growth in the building sectors rather than the existing traditional concretes in terms of enhanced environmental friendliness and pollution reduction.

1.2   CEMENT-GENERATED ENVIRONMENTAL PROBLEMS

Universally, OPC has been continually exploited as a concrete binder and with different building substances. It is known that production of OPC-based concretes needs huge amount of fuel and raw ingredients that are acquired via resource mining and energy exhaustive processing [11]. This in turn causes great quantities of greenhouse gases (essentially CO_2 and NO_x) emission into the atmosphere. Per ton of OPC manufacturing, almost one ton of CO_2 is created by consuming 2.5 tons of raw materials and fuel [22–25]. Some estimates reveal that approximately 1.35 billion tons of CO_2 is emitted per annum from the OPC production industries alone, accounting for nearly 7% of the total greenhouse gases released to the environment [10, 26]. Compared to CO_2 levels in 2010, the CO_2 emissions from OPC industries on the earth's atmosphere is going to be a major environmental concern because the global demand of OPC is predicted to grow nearly 200% by 2050 [10]. Over the years, numerous strategies were adopted to reduce OPC manufacturing, including the durability enhancement of concrete and elongation of service lifespan. From this standpoint, self-healing concrete has emerged as one of the gifted solutions to mitigate the OPC-generated environmental pollution.

1.3   ENERGY PROBLEMS IN CEMENT INDUSTRIES

Nowadays, cement is the primary workhorse in the construction sectors worldwide. Cement is processed at high temperature ($\approx$1450°C) in a rotary kiln by mixing limestone or chalks with clay. Next, the yielded hard nodules of clinker are crushed with low amount of gypsum using a ball milling technique. Firing of such cement constituents at higher temperature needs substantial amounts of energy to be consumed (burning of coal or petroleum coke). Rapid decay of land setting, generation of dust during transport, creation of noise throughout quarrying, and processing of raw materials are regarded as major environmental concerns in OPC production. According to previous publications [27–28], OPC manufacturing devoid of considerable amounts of CO_2 release is impossible. Actually, the CO_2 emission occurs in two phases, including fuel burning (to achieve very high temperatures of kiln) and calcining (a chemical reaction occurs during limestone firing). Until now, even the most efficient cement production plants emit 60% or more of CO_2 from various inevitable chemical pathways. The consumption of a considerable quantity of energy during crushing of cementitious raw materials in the clinker phase remains the foremost ecological distress in the OPC industries [29–30]. Undeniably, developing autonomic self-healing cement-based materials must be realized as a practical solution to the existing problems.

1.4 CONCRETE PERFORMANCE IN AGGRESSIVE ENVIRONMENTS

Economically, the building materials serviceability is of great importance, especially in the contemporary infrastructures and constituent parts. In the context of the urban development, the most commonly employed concrete materials have to comply with the requirements of the standard codes of practice with regard to strength and durability [31–33]. For instance, the serviceability of the concrete material in use can be reduced by factors such as suboptimal planning, low capacity, or overload; flaws in the material design and structures; incorrect building practices or substandard maintenance; or a lack of knowledge about the engineering [34–35]. The concrete structures used in the construction industries require an extra improvement because they deteriorate rapidly during their lifespans. This deterioration is due to a range of both extrinsic and intrinsic processes related to chemical, physical, thermal, and biological nature [36]. Furthermore, the impact of the inadequate use and environmental conditions on how concrete performs has been highlighted in many studies [28, 37–39]. Within the construction industries, the prevention of the steel reinforcement exposure to hazardous chemicals such as the corrosive agents depends significantly on the concrete. In general, the steel reinforcement becomes exposed to the corrosion as a result of the crack formation in the concrete, enabling harsh chemicals (e.g. chloride) to penetrate and reach steel reinforcement bar. After reacting with water and oxygen, these chemicals produce corrosion [40]. Figure 1.1 illustrates a basic representation of the occurrence of the corrosion in the reinforced concrete.

Cracks are detrimental not only in terms of facilitating corrosion, but also concerning the aesthetics because they make the visible porous structure in the concrete, expanding in size if no remedial action is taken. Hazardous chemicals can permeate the concrete through the cracks of large size, leading to the chemical or physical deterioration of the concrete. However, it is not possible to completely prevent the micro-cracks in the concrete because it is too expensive in terms of maintenance and repairs [40]. Consequently, additional funding allocation is needed for the maintenance work with regard to the necessary materials and skilled workers. Based on these factors, it is realized that the materials with self-repair capability can make a

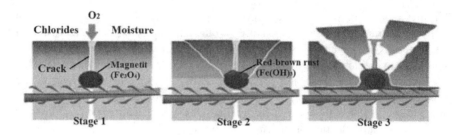

FIGURE 1.1   Process of aggressive solution penetration [16].

substantial difference by repairing the cracks automatically and thus not only reduce the expenditure, but also increase the lifespan of the concrete structures.

1.5   CRACK PROBLEMS IN CONCRETE

The formation of cracks in concrete is a major issue that must be addressed adequately. Drying-related shrinkage, thermal contraction, external or internal restraint to shortening, subgrade settlement, and overloading can all determine the formation of the cracks. Although complete avoidance of crack formation is nearly impossible, some methods are available for mitigating such issues. The main points at which the cracks form are prior to and following the concrete hardening [41]. In order to prevent the future degradation that can reduce the use life of the concrete structure, knowledge must be acquired about the causes and remedial measures to be adopted in relation to the crack formation at those crucial moments.

Settlement of concrete is the key determinant of the pre-hardening cracks formation. It starts when water is lost in the plastic state. The factors including the lack of adequate vibration, high slumps associated with the exceedingly wet concretes, or insufficient covering of the embedded items (for example, steel reinforcement) or at the margin of the concrete are responsible for the settlement cracking. In addition, the plastic shrinking can engender the pre-hardening cracking [42]. Such cracks usually occur in the slabs prior to the final finishing and under various environmental conditions, including the strong wind, low humidity, and high daytime temperatures. These conditions promote a rapid evaporation of the moisture from the surface, and thus determining a greater surface shrinkage compared to the interior of the concrete [43]. The surface shrinkage is restrained by the interior concrete, leading to the occurrence of the stresses higher than the tensile strength of the concrete, which in turn leads to the formation of the cracks at the surface. The length of the cracks related to the plastic shrinkage is variable, but they frequently reach to the slab mid-depth. Such cracks can be attenuated through the method of fogging that is generally implemented at the construction site.

The factors that can stimulate the crack formation and post-hardening of the concrete include drying-related shrinkage, thermal contraction, and subgrade settlement. The strategy of inserting regularly spaced construction joints is usually applied to avoid shrinkage and manage crack location. For instance, the joints can be inserted in such a way as to determine the formation of the cracks in locations where they can be anticipated without difficulty. Furthermore, the number of cracks can be minimized through the introduction of horizontal steel reinforcement that can further hinder excessive crack expansion.

1.6 SUSTAINABILITY OF SMART CONCRETE

Low carbon emission and energy-saving building material incorporated with smart materials (in self-healing technology) is a well-known candidate energy technology in enhancing energy efficiency and sustainability of building. The major aim of sustainable development is to keep livelihood on earth in the predictable future with absolute support or care so that ecological balance is not disturbed [44]. Sustainability is founded on three basic elements: economic security, environmental safety, and societal benefits. Sustainable advancement must preserve these factors to protect the biodiversity within a balanced ecosystem. In the present industrial uprising era, engineers, scientists, policymakers, and architects are attempting to use the sustainable model resourcefully to reduce negative impact on our ecosystem. Therefore, in the perspective of building materials, the term "sustainability" is used synonymously with robust or friendly and green environment [45–46]. In this viewpoint, self-healing materials have attracted increasing interest due to their potential to lessen degradation, prolong functional lifespan, and suppress maintenance costs [47–48]. However, the self-healing technology contributes directly to enhancement of the environment reduction of pollution from increasing concrete lifespan and reducing the demand and consumption of OPC, as well as affecting energy savings and increasing the sustainability of concrete.

1.7 EFFECT OF ADDITION SELF-HEALING AGENTS ON CONCRETE PROPERTIES

It is established that the inclusion of self-healing materials in the concrete matrix can offer several advantages and disadvantages in the mechanical properties, depending on the nature of the materials and the self-healing process. It was found that the insertion of capsules in the concrete matrix can result in voids or holes in the concrete after releasing the contained agent. These voids in turn can affect negatively the strength performance of the concrete [20]. The use of the bacteria spores as the self-healing agent in the concrete was shown to reduce the compressive strength performance with loss in strength from 8–10%. In addition, the loss in strength increases with the increase in the bacteria dosages [49]. The observed strength reduction was attributed to the microstructures alteration induced by the reduced degree of hydration and poor distribution of the hydration products because of the nutrients and microcapsules inclusion. Algaifi et al. [50] reported that the inclusion of the microbial calcium carbonate in the self-healing concrete matrix can improve the compressive strength when compared with the normal concrete specimens. Similar results were obtained by Shaheen et al. [51], wherein the mechanical properties of the prepared concrete— such as the compressive strength and splitting tensile strength—were enhanced when immobilization techniques for the self-healing process were utilized.

1.8   LIFE CYCLE ANALYSIS OF SELF-HEALING CONCRETE

Ample research has been conducted on the self-repair technology in the last ten years, and its potential for promoting an autonomous crack healing in the concrete has been highlighted. This has led to the development of several self-repair mechanisms for cementitious materials. The evaluation of various products and services in terms of their effect on the environment from the moment of development until discontinuation is based on the life cycle assessment (LCA) methodology achieved the standardization as ISO 14040–14044. In essence, the LCA is geared toward determining whether the self-repairing concrete is more sustainable than standard concrete. Diminished deterioration rate, prolonged use life, and decreased frequency and low cost of repairs over the lifespan of a concrete infrastructure are among the main advantages of self-repairing concrete. These can make the concrete more sustainable, as reduced repair frequency translates to a decrease in the amount of material resources and energy used, a decrease in the environmentally damaging emissions associated with the manufacture and transport of materials, and a reduction in the traffic changes to transport infrastructure required by the repair or reconstruction work [52–53].

Van et al. [54] discovered that the chlorides can be prevented from instantly permeating through cracks when encapsulated polyurethane precursor was used as self-repair material. Furthermore, the levels of chloride in an area with the cracks was decreased by at least 75% in the self-repairing concrete. Compared to the standard concrete that has a use life in marine environments of just seven years, the concrete with self-repair capability is usable in such environments for about 60–94 years. The lengthening of the lifespan was the main determinant of the significant environmental advantages (56–75%), according to the computations of life cycle evaluation.

1.9   SUMMARY

A comprehensive overview of the relevant literature on cementitious materials sustainability enabled us to draw the following conclusions:

    i. Perhaps the greatest challenge in fighting climate change comes from the production of cement, which is the main constituent of concrete—and the world's appetite for it seems insatiable.

    ii. Environmentally sustainable construction materials with reduced carbon footprint have ever-growing demand worldwide.

    iii. The concrete structures used in the construction industry require extra improvement because they deteriorate rapidly during their lifespans. This deterioration is due to a range of both extrinsic and intrinsic processes related to chemical, physical, thermal, and biological nature.

    iv. Rapid deterioration of conventional concrete in aggressive environments is a major problem in the concrete industry.

    v. It is established that the inclusion of self-healing materials in the concrete matrix can offer several advantages in its durability properties, depending on the nature of the materials and the self-healing process.

REFERENCES

1. Kubba, Z., et al., *Impact of curing temperatures and alkaline activators on compressive strength and porosity of ternary blended geopolymer mortars.* Case Studies in Construction Materials, 2018.**9**: p. e00205.

2. Samadi, M., et al., *Waste ceramic as low cost and eco-friendly materials in the production of sustainable mortars.* Journal of Cleaner Production, 2020.**266**: p. 121825.

3. Chen, R., et al., *Effect of particle size of fly ash on the properties of lightweight insulation materials.* Construction and Building Materials, 2016.**123**: p. 120–126.

4. Huseien, G.F., et al., *Waste ceramic powder incorporated alkali activated mortars exposed to elevated temperatures: Performance evaluation.* Construction and Building Materials, 2018.**187**: p. 307–317.

5. Mohammadhosseini, H., M.M. Tahir, and M. Sayyed, *Strength and transport properties of concrete composites incorporating waste carpet fibers and palm oil fuel ash.* Journal of Building Engineering, 2018.**20**: p. 156–165.

6. Huseien, G.F., et al., *The effect of sodium hydroxide molarity and other parameters on water absorption of geopolymer mortars.* Indian Journal of Science and Technology, 2016.**9**(48).

7. Du, K., C. Xie, and X. Ouyang, *A comparison of carbon dioxide (CO 2) emission trends among provinces in China.* Renewable and Sustainable Energy Reviews, 2017.**73**: p. 19–25.

8. Keyvanfar, A., et al., *User satisfaction adaptive behaviors for assessing energy efficient building indoor cooling and lighting environment.* Renewable and Sustainable Energy Reviews, 2014.**39**: p. 277–295.

9. Huseien, G.F., et al., *Geopolymer mortars as sustainable repair material: A comprehensive review.* Renewable and Sustainable Energy Reviews, 2017.**80**: p. 54–74.

10. Xie, T., and T. Ozbakkaloglu, *Behavior of low-calcium fly and bottom ash-based geopolymer concrete cured at ambient temperature.* Ceramics International, 2015.**41**(4): p. 5945–5958.

11. Huseien, G.F., et al., *Influence of different curing temperatures and alkali activators on properties of GBFS geopolymer mortars containing fly ash and palm-oil fuel ash.* Construction and Building Materials, 2016.**125**: p. 1229–1240.

12. Palomo, Á., et al., *Railway sleepers made of alkali activated fly ash concrete.* Revista Ingeniería de Construcción, 2011.**22**(2): p. 75–80.

13. Huseiena, G.F., et al., *Potential use coconut milk as alternative to alkali solution for geopolymer production.* Jurnal Teknologi, 2016.**78**(11): p. 133–139.

14. Ariffin, N.F., et al., *Strength properties and molecular composition of epoxy-modified mortars.* Construction and Building Materials, 2015.**94**: p. 315–322.

15. Huseien, G.F., et al., *Synthesis and characterization of self-healing mortar with modified strength.* Jurnal Teknologi, 2015.**76**(1).

16. Shah, K.W., and G.F. Huseien, *Biomimetic self-healing cementitious construction materials for smart buildings.* Biomimetics, 2020.**5**(4): p. 47.

17. Jiang, L., and D. Niu, *Study of deterioration of concrete exposed to different types of sulfate solutions under drying-wetting cycles.* Construction and Building Materials, 2016.**117**: p. 88–98.

18. Chen, Y., et al., *Resistance of concrete against combined attack of chloride and sulfate under drying—wetting cycles.* Construction and Building Materials, 2016.**106**: p. 650–658.

19. Calvo, J.G., et al., *Development of ultra-high performance concretes with self-healing micro/nano-additions.* Construction and Building Materials, 2017.**138**: p. 306–315.

20. Gupta, S., S. Dai Pang, and H.W. Kua, *Autonomous healing in concrete by bio-based healing agents—A review.* Construction and Building Materials, 2017.**146**: p. 419–428.

21. Wang, J., et al., *Use of silica gel or polyurethane immobilized bacteria for self-healing concrete.* Construction and Building Materials, 2012.**26**(1): p. 532–540.

22. Miranda, J., et al., *Corrosion resistance in activated fly ash mortars.* Cement and Concrete Research, 2005.**35**(6): p. 1210–1217.

23. Habert, G., J.D.E. De Lacaillerie, and N. Roussel, *An environmental evaluation of geopolymer based concrete production: Reviewing current research trends.* Journal of Cleaner Production, 2011.**19**(11): p. 1229–1238.

24. Huseien, G.F., et al., *Effect of metakaolin replaced granulated blast furnace slag on fresh and early strength properties of geopolymer mortar.* Ain Shams Engineering Journal, 2018.**9**: p. 1557–1566.

25. Mohammadhosseini, H., et al., *Durability performance of green concrete composites containing waste carpet fibers and palm oil fuel ash.* Journal of Cleaner Production, 2017.**144**: p. 448–458.

26. Shi, X., et al., *Mechanical properties and microstructure analysis of fly ash geopolymeric recycled concrete.* Journal of Hazardous Materials, 2012.**237**: p. 20–29.

27. Gartner, E., *Potential improvements in cement sustainability.* 31st cement and concrete science conference novel developments and innovation in cementitious materials, 12–13 September 2011.

28. Huseien, G.F., et al., *Effects of POFA replaced with FA on durability properties of GBFS included alkali activated mortars.* Construction and Building Materials, 2018.**175**: p. 174–186.

29. Li, J., S.T. Ng, and M. Skitmore, *Review of low-carbon refurbishment solutions for residential buildings with particular reference to multi-story buildings in Hong Kong.* Renewable and Sustainable Energy Reviews, 2017.**73**: p. 393–407.

30. Wang, P., et al., *Life cycle assessment of magnetized fly-ash compound fertilizer production: A case study in China.* Renewable and Sustainable Energy Reviews, 2017.**73**: p. 706–713.

31. Behfarnia, K., *Studying the effect of freeze and thaw cycles on bond strength of concrete repair materials.* Asian Journal of Civil Engineering, 2010.**11**(2): p. 165–172.

32. Kubba, Z., et al., *Effect of sodium silicate content on setting time and mechanical properties of multi blend geopolymer mortars.* Journal of Engineering and Applied Science, 2019.**14**: p. 2262–2267.

33. Huseien, G.F., et al., *Evaluation of alkali-activated mortars containing high volume waste ceramic powder and fly ash replacing GBFS.* Construction and Building Materials, 2019.**210**: p. 78–92.

34. Gouny, F., et al., *A geopolymer mortar for wood and earth structures.* Construction and Building Materials, 2012.**36**: p. 188–195.

35. Huseien, G.F., and K.W. Shah, *Performance evaluation of alkali-activated mortars containing industrial wastes as surface repair materials.* Journal of Building Engineering, 2020.**30**: p. 101234.

36. Mirza, J., et al., *Preferred test methods to select suitable surface repair materials in severe climates.* Construction and Building Materials, 2014.**50**: p. 692–698.

37. Mueller, H.S., et al., *Design, material properties and structural performance of sustainable concrete.* Procedia Engineering, 2017.**171**: p. 22–32.

38. Qureshi, T., A. Kanellopoulos, and A. Al-Tabbaa, *Autogenous self-healing of cement with expansive minerals-II: Impact of age and the role of optimised expansive minerals in healing performance.* Construction and Building Materials, 2019.**194**: p. 266–275.

39. Huseien, G.F., K.W. Shah, and A.R.M. Sam, *Sustainability of nanomaterials based self-healing concrete: An all-inclusive insight.* Journal of Building Engineering, 2019.**23**: p. 155–171.

40. Al-Zahrani, M., et al., *Effect of waterproofing coatings on steel reinforcement corrosion and physical properties of concrete.* Cement and Concrete Composites, 2002.**24**(1): p. 127–137.

41. Sakulich, A.R., L. Kan, and V.C. Li, *Microanalysis of autogenous healing products in engineered cementitious composites (ECC).* Microscopy and Microanalysis, 2010.**16**(S2): p. 1220–1221.

42. Qian, S., et al., Influence of microfiber additive effect on the self-healing behavior of engineered cementitious composites. In *Sustainable construction materials, Wuhan, China, 2012.* 2013: ASCE. p. 203–214.

43. Zhong, W., and W. Yao, *Influence of damage degree on self-healing of concrete.* Construction and Building Materials, 2008.**22**(6): p. 1137–1142.

44. Mastrucci, A., et al., *Life cycle assessment of building stocks from urban to transnational scales: A review.* Renewable and Sustainable Energy Reviews, 2017.**74**: p. 316–332.

45. Struble, L., and J. Godfrey. *How sustainable is concrete.* International workshop on sustainable development and concrete technology, 2004.

46. Bilodeau, A., and V.M. Malhotra. *High-volume fly ash system: The concrete solution for sustainable development.* CANMET/ACI. Séminaire International, 2000.

47. Zhu, D.Y., M.Z. Rong, and M.Q. Zhang, *Self-healing polymeric materials based on microencapsulated healing agents: From design to preparation.* Progress in Polymer Science, 2015.**49**: p. 175–220.

48. He, Z., et al., *Facile and cost-effective synthesis of isocyanate microcapsules via polyvinyl alcohol-mediated interfacial polymerization and their application in self-healing materials.* Composites Science and Technology, 2017.**138**: p. 15–23.

49. Jonkers, H.M., et al., *Application of bacteria as self-healing agent for the development of sustainable concrete.* Ecological Engineering, 2010.**36**(2): p. 230–235.

50. Algaifi, H.A., et al., *Insight into the role of microbial calcium carbonate and the factors involved in self-healing concrete.* Construction and Building Materials, 2020.**254**: p. 119258.

51. Shaheen, N., et al., *Synthesis and characterization of bio-immobilized nano/micro inert and reactive additives for feasibility investigation in self-healing concrete.* Construction and Building Materials, 2019.**226**: p. 492–506.

52. Li, V.C., and E. Herbert, *Robust self-healing concrete for sustainable infrastructure.* Journal of Advanced Concrete Technology, 2012.**10**(6): p. 207–218.

53. Huseien, G.F., et al., *Utilizing spend garnets as sand replacement in alkali-activated mortars containing fly ash and GBFS.* Construction and Building Materials, 2019.**225**: p. 132–145.

54. Van Belleghem, B., et al., *Quantification of the service life extension and environmental benefit of chloride exposed self-healing concrete.* Materials, 2017.**10**(1): p. 5.

2 Self-Healing Technology
Fundamental and Design Strategies

2.1 INTRODUCTION

Human settlements are dependent on infrastructure systems (e.g. buildings, transport, energy, water, communication) to mediate a wide range of human activities [1–2]. Globally, the infrastructures made of concrete (e.g. bridges, buildings, wharves, etc.) are susceptible to deterioration due to the ever-increasing carbon dioxide levels, temperatures, and relative humidity [3]. Concrete is primarily made up of cement, the production of which has major implications for climate change, especially it appears that the demand for cement is relentless. The concrete structure durability and lifespan under different weather conditions are influenced greatly by the high sensitivity of the cement materials to harsh environmental conditions that can cause concrete expansion and eventual cracking [4–5]. In turn, the severe structural issues can arise from the concrete cracking. Thus, smart concrete based on self-repair technology and exceptionally durability has been developed in response to the ever-greater specifications regarding the durability that must be satisfied by these concrete structures [6].

The impact of climate change on human life and ecosystems can be varied and manifest either in a direct or indirect manner. These include rising sea levels, intensification in the rate and/or severity of extreme weather phenomena, increasing frequency of heat waves, and enhanced rates of precipitation. Thus, from the perspective of engineering, more and more research attention must be directed toward the effect of climate change on the structural features of the concretes. Simultaneously, risk-based techniques can effectively be employed to evaluate how the viable strategies of the climate adaptation produce an economic benefit. For example, a model for improving the hurricane risk evaluation was developed by Bjarnadottir et al. [7] by taking into account the effect of climate change on the intensity and/or frequency of hurricanes. Meanwhile, the stochastic method suggested by Bastidas-Arteaga et al. [8] for the concrete structures was intended to facilitate the investigation of the global warming impact on chloride ingress, although the study was limited to the incipient phase of the corrosion. In a different study, Stewart and Peng [9] undertook a preliminary risk and cost-benefit assessment related to the adaptation measures intended to counteract the impact of the carbonation of the reinforced concrete structures (RCs). The effect of climate change on the concrete structure durability in various regions has also been explored in many recent studies. For instance, the contribution of climate change to concrete deterioration triggered by corrosion was examined by Stewart et al. [10] and Wang et al. [11]. They considered a probabilistic method for

DOI: 10.1201/9781003195764-2

evaluating the corrosion-related cracking and spalling related to the effect of climate change on regions with distinct geographical conditions. Additionally, the impact of the climate change on the carbonation in Toronto and Vancouver (Canada) was the focus of the study by Talukdar et al. [12], wherein the carbonation depths were found to increase by around 45% over a period of a century. On the downside, the study disregarded the ambiguities associated with the climate, materials, and models, and failed to estimate the extent to which the concrete infrastructure deterioration and safety were influenced by the carbonation.

The durability of a material is typically aided by the self-repair capability, especially in the challenging situations for humans to intervene like construction situated in areas with deleterious physical and chemical conditions [13–15]. Another use of self-repair is the protection of the material properties, particularly in the kinetic and thermodynamic conditions supporting large defects density like nanostructures. Nanomaterials always have superior functional characteristics, and unlike traditional materials, they display a quicker rate of the deterioration because of the high content of the interfacial atoms. A wide range of nanosystems can be created through the integration of numerous functional nanostructures with some constituents added to provide the self-repairing capability. Indeed, by comparison to the development of a nanosystem with greater robustness, such an approach is quite basic [16]. Rapid innovations in nanoscience and nanotechnology have recently revolutionized the development of the materials with self-repair capability, including nanomaterials, which consist of particles of less than 500 nm in size. In such materials, the degeneration recovery occurs without external intervention with self-repair capability. However, the external stimuli (e.g. temperature) can activate the self-repair, as well, which is the case with systems known as non-autonomic self-healing materials [17].

The purpose of the present chapter is to review the current investigations on the concrete structures based on various design strategies with the self-repair capability and their implications for the future uses in the sustainable projects. In short, this presentation consists of three parts. First, it is argued in Section 2.2 that unlike standard concretes, self-repairing concretes are more eco-friendly and can reduce pollution levels, thereby enabling the construction industry to move toward greater sustainability and intelligent development. Section 2.3 deals with mechanisms of self-healing in cementitious materials. Section 2.4 involves the self-repairing systems that are proposed as viable options of the concrete self-recovery mechanisms in response to corrosion, damage, and cracking.

2.2   SUSTAINABILITY OF SELF-HEALING CONCRETE

Self-repair technology is a new innovation in the concrete industry that refers to the materials having high quality and which are able to repair damage on their own without any interference from the outside. In fact, this technology is introduced in order to satisfy the demands for reduction in the expenditures related to concrete structural repair and maintenance [16]. Intense interest has been generated by this technology in the last ten years, due to its potential use in the building structures. Self-repair technology is also known as autonomic healing, autonomic repair, and self-healing [18]. The main applications of this technology at the moment are repair of cracks to restore mechanical strength and automatic crack repair to avoid extra financial expenditures and the necessity for additional raw materials [19].

To make the concrete structures more serviceable and expand their use life, crack repair is essential. A material with self-repair capability is referred to as intelligent or smart material, meaning that it systematically integrates both the construct of information and physical indices like strength and durability [20], affording the material functionality at a higher level. Thus, a smart material is able to regulate itself, having a capacity for the detection and response, as well as controlled delivery of the response. The innate ability of the natural materials and their mechanical properties to adapt intelligently has been explored in earlier studies [21]. Conversely, human-made smart materials are yet to fully mature and their applications are restricted to medicine, bionics, and aeronautics or astronautics.

Figure 2.1 shows a schematic diagram of the main mechanisms of self-healing concrete. Depicted are the formation of the calcium carbonate from calcium hydroxide, settlement of the debris and loose cement particles in presence of water, hydration of unhydrated cementitious particles, and further swelling of the hydrated cementitious matrix [22]. It is shown that a different self-healing reaction can occur depending on the self-healing agent used in the concrete. For instance, the use of bacteria as the self-healing agent can lead to the generation of calcium carbonate because of the chemical reactions among the bacteria, oxygen, and water. The bacteria produce a calcite participate via these reactions. Other expansive self-healing agents can fill up the cracks via the swelling of the materials, and thus repair the concrete automatically.

One type of energy technology considered to have great potential in making the building of structures more sustainable and energy efficient is a low carbon emission and energy-efficient building material with self-repair capability. Sustainable development is geared toward the careful management of human activities to ensure the future survival of humanity on this planet while avoiding any disruption to the ecological equilibrium [23]. Therefore, economic security, environmental safety, and societal benefits are the pillars of sustainability that have to be upheld for safeguarding of biodiversity and balancing of ecosystems. Within the current industry-dominated world, dedicated efforts are being made in numerous fields (e.g.

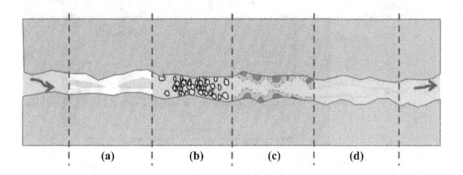

FIGURE 2.1 Main mechanisms of autogenous healing [22].

engineering, science, policy-making, architecture) to achieve the resourceful implementation of sustainability to attenuate the adverse effects of human activities on
the ecosystems. Regarding the construction materials, sustainability implies the low
environmental impact of the materials [24–25]. In this context, the growing attention
is being paid to the self-repair technology because it can diminish the deterioration
of material, expand useful life, and eliminate maintenance expenditures [26–27].
Hence, self-repair technology can make construction materials and concrete more
sustainable by lengthening their use life and decreasing demand for and consumption
of OPC, improving energy efficiency and lowering pollution levels.

2.3 MECHANISMS OF SELF-HEALING IN CEMENTITIOUS MATERIALS

In the human body, skin and tissues are capable of repairing themselves through
the replacement of the damaged areas based on nutrient uptake. Similarly, essential
products serve as nutrients in the cement-based materials with the self-repair capability, and may enable these materials to repair the damage or deterioration (Figure 2.2).
Ample research has been conducted in recent times to discover the new methods for
achieving an effective self-repair material alongside the durability of the cement-
based materials. Figure 2.3 illustrates an overview of such methods.

2.4 DESIGN STRATEGIES

The different types of materials such as plastics/polymers, paints/coatings, metals/
alloys, and ceramics/concrete have their own self-healing mechanisms. Here, different types of self-healing processes are discussed with respect to design strategies and
not with respect to types of materials and their related self-healing mechanisms. The
self-healing processes are:

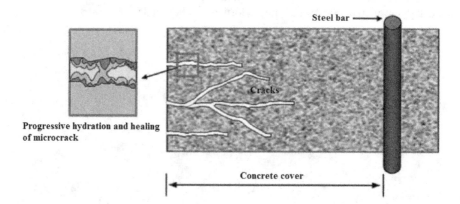

FIGURE 2.2   Healing of micro-cracks in concrete cover due to continuing hydration of
unhydrated cement nuclei [28].

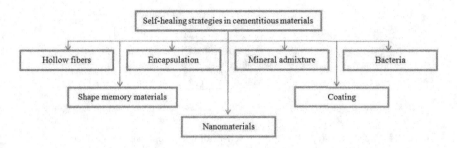

FIGURE 2.3 Developed strategies for self-healing in cement-based materials.

- Release of healing agent
- Reversible cross-links
- Miscellaneous technologies

2.4.1 RELEASE OF HEALING AGENT

Liquid active agents such as monomers, dyes, catalysts, and hardeners containing microcapsules, hollow fibers, or channels are embedded into polymeric systems during its manufacturing stage. In the case of a crack, these reservoirs are ruptured and the reactive agents are poured into the cracks by capillary force, whereby it solidifies in the presence of predisposed catalysts and heals the crack. The propagation of cracks is the major driving force of this process. On the other hand, it requires the stress from the crack to be relieved, which is a major drawback of this process. As this process does not need a manual or external intervention, it is autonomic. The remainder of this subsection gives an overview of different possibilities to explore this concept of designing self-healing materials. The possible concepts are:

- Microcapsule embedment
- Hollow fiber embedment
- Microvascular system

Microencapsulation is a process of enclosing micron-sized particles of solids, droplets of liquids, or gases in an inert shell, which in turn isolates and protects them from the external environments [29]. The end product of the microencapsulation process is termed as microcapsules. They may have spherical or irregular shapes, and may vary in size ranging from nano to micro scale. Healing agents or catalysts containing microcapsules are used to design self-healing polymer composites. Early literature [30] suggests the use of microencapsulated healing agents in a polyester matrix to achieve a self-healing effect, but they were unsuccessful in producing practical self-healing materials. The first practical demonstration of self-healing materials was performed in 2001 by Prof. Scot White and his collaborators [31]. Self-healing capabilities were achieved by embedding encapsulated healing agents into polymer matrix containing dispersed catalysts. The self-healing strategy used by them is shown in Figure 2.4.

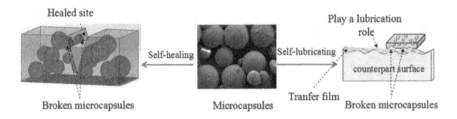

FIGURE 2.4 Schematic representation of self-healing concept using embedded microcapsules [32].

Microcapsule-based self-healing approach has the major disadvantage of uncertainty in achieving complete and/or multiple healing as it has limited amount of healing agent. Multiple healing is only feasible when excess healing agent is available in the matrix after the first healing has occurred. Thus, to achieve multiple healing in composite materials, another type of reservoir that might be able to deliver larger amount of liquid healing agent was developed. However, they have achieved only limited success using their approach. Composite systems formulated on the basis of these filled glass fibers were unable to deliver the resin into the crack owing to the use of high viscous epoxy resins, and curing was also not good (Figure 2.5).

To overcome the difficulty of short supply of a healing agent in microcapsule-based self-healing concept, another approach similar to biological vascular system of many plants and animals was explored by Therriault et al. [34]. This approach relies on a centralized network (that is, a microvascular network) for distribution of healing agents into polymeric systems in a continuous pathway. The fabrication process is complex, and it is very difficult to achieve synthetic materials with such networks for practical applications. In this process, organic inks are deposited following a 3D array and the interstitial pores between the printed lines are infiltrated with an epoxy resin. Once the polymer is cured, the fugitive ink is removed leaving behind a 3D microvascular channel with well-defined connectivity. Polymeric systems with microvascular networks were prepared by incorporating chemical catalysts in the polymer used to infiltrate the organic ink scaffold. Upon curing the polymer and removing the scaffold, the healing agent is wicked into the microvascular channels. Several researchers reported such fabrication processes and related self-healing capabilities [35].

2.4.2 Reversible Cross-Links

Cross-linking of polymeric materials, which is an irreversible process, is performed to achieve superior mechanical properties, such as high modulus, solvent resistance, and high fracture strength. However, it adversely affects the refabrication ability of polymers. Moreover, highly cross-linked materials have the disadvantage of brittleness and have the tendency to crack. One approach to bring process ability to cross-linked polymers is the introduction of reversible cross-links in polymeric systems. In addition to refabrication and recyclability, reversible cross-links also exhibit self-healing properties. However, reversible cross-linked systems do not show

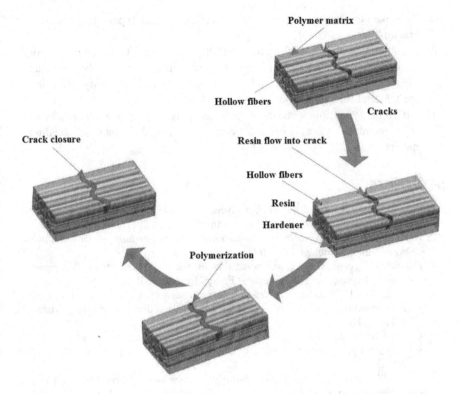

FIGURE 2.5 Schematic representation of self-healing concept using hollow fibers [33].

self-repairing ability on their own. An external trigger such as thermal, photo, or chemical activation is needed to achieve reversibility, and thereby the self-healing ability. Thus, these systems show non-autonomic healing phenomenon. Later in this book, different approaches that are considered to bring reversibility in cross-linked polymeric materials are discussed [35].

2.4.3 MISCELLANEOUS TECHNOLOGIES

Technologies other than the most important self-healing approaches described previously are available in the literature. These emerging technologies—electro hydrodynamics, conductivity, shape memory effect, nanoparticle migrations and co-deposition [35]—are discussed later in this book.

2.5 SUMMARY

From the reviewed studies, the several conclusions can be made:

i. Concrete structure durability and lifespan under different weather conditions are influenced greatly by the high sensitivity of the cement materials to harsh environmental conditions that can cause concrete expansion and eventual cracking.

ii. To make the concrete structures more serviceable and expand their use life, crack repair is essential.

iii. A material with the self-repair capability is referred to as intelligent or smart material, meaning that it systematically integrates both the construct of information and physical indices like strength and durability, affording the material functionality at a higher level.

iv. Different types of materials—such as plastics/polymers, paints/coatings, metals/alloys, and ceramics/concrete—have their own self-healing mechanisms.

REFERENCES

1. Stewart, M.G., X. Wang, and M.N. Nguyen, *Climate change impact and risks of concrete infrastructure deterioration.* Engineering Structures, 2011.**33**(4): p. 1326–1337.

2. Huseien, G.F., et al., *Synthesis and characterization of self-healing mortar with modified strength.* Jurnal Teknologi, 2015.**76**(1).

3. Huseien, G.F., et al., *Geopolymer mortars as sustainable repair material: A comprehensive review.* Renewable and Sustainable Energy Reviews, 2017.**80**: p. 54–74.

4. Jiang, L., and D. Niu, *Study of deterioration of concrete exposed to different types of sulfate solutions under drying-wetting cycles.* Construction and Building Materials, 2016.**117**: p. 88–98.

5. Chen, Y., et al., *Resistance of concrete against combined attack of chloride and sulfate under drying—wetting cycles.* Construction and Building Materials, 2016.**106**: p. 650–658.

6. Calvo, J.G., et al., *Development of ultra-high performance concretes with self-healing micro/nano-additions.* Construction and Building Materials, 2017.**138**: p. 306–315.

7. Bjarnadottir, S., Y. Li, and M.G. Stewart, *A probabilistic-based framework for impact and adaptation assessment of climate change on hurricane damage risks and costs.* Structural Safety, 2011.**33**(3): p. 173–185.

8. Bastidas-Arteaga, E., et al., *Influence of weather and global warming in chloride ingress into concrete: A stochastic approach.* Structural Safety, 2010.**32**(4): p. 238–249.

9. Stewart, M.G., and J. Peng, *Life cycle cost assessment of climate change adaptation measures to minimise carbonation-induced corrosion risks.* International Journal of Engineering Under Uncertainty: Hazards, Assessment and Mitigation, 2010.**2**(1–2): p. 35–46.

10. Stewart, M.G., X. Wang, and M.N. Nguyen, *Climate change adaptation for corrosion control of concrete infrastructure.* Structural Safety, 2012.**35**: p. 29–39.

11. Wang, X., M.G. Stewart, and M. Nguyen, *Impact of climate change on corrosion and damage to concrete infrastructure in Australia.* Climatic Change, 2012.**110**(3): p. 941–957.

12. Talukdar, S., et al., *Carbonation in concrete infrastructure in the context of global climate change: Part 2—Canadian urban simulations.* Cement and Concrete Composites, 2012.**34**(8): p. 931–935.

13. Samadi, M., et al., *Influence of glass silica waste nano powder on the mechanical and microstructure properties of alkali-activated mortars.* Nanomaterials, 2020.**10**(2): p. 324.

14. Samadi, M., et al., *Waste ceramic as low cost and eco-friendly materials in the production of sustainable mortars.* Journal of Cleaner Production, 2020.**266**: p. 121825.

15. Huseien, G.F., and K.W. Shah, *Durability and life cycle evaluation of self-compacting concrete containing fly ash as GBFS replacement with alkali activation.* Construction and Building Materials, 2020.**235**: p. 117458.

16. Gupta, S., S. Dai Pang, and H.W. Kua, *Autonomous healing in concrete by bio-based healing agents—A review*. Construction and Building Materials, 2017.**146**: p. 419–428.

17. Wang, J., et al., *Use of silica gel or polyurethane immobilized bacteria for self-healing concrete*. Construction and Building Materials, 2012.**26**(1): p. 532–540.

18. Mangun, C., et al., *Self-healing of a high temperature cured epoxy using poly (dimethyl-siloxane) chemistry*. Polymer, 2010.**51**(18): p. 4063–4068.

19. Thao, T.D.P., et al., *Implementation of self-healing in concrete—Proof of concept*. The IES Journal Part A: Civil & Structural Engineering, 2009.**2**(2): p. 116–125.

20. Mihashi, H., and T. Nishiwaki, *Development of engineered self-healing and self-repairing concrete-state-of-the-art report*. Journal of Advanced Concrete Technology, 2012.**10**(5): p. 170–184.

21. Zwaag, S., *Self healing materials: An alternative approach to 20 centuries of materials science*. Vol. 30. 2008: Springer Science+ Business Media BV Dordrecht, The Netherlands.

22. Tang, W., O. Kardani, and H. Cui, *Robust evaluation of self-healing efficiency in cementitious materials—A review*. Construction and Building Materials, 2015.**81**: p. 233–247.

23. Mastrucci, A., et al., *Life Cycle Assessment of building stocks from urban to transnational scales: A review*. Renewable and Sustainable Energy Reviews, 2017.**74**: p. 316–332.

24. Struble, L., and J. Godfrey. *How sustainable is concrete*. International workshop on sustainable development and concrete technology, 2004.

25. Bilodeau, A., and V.M. Malhotra, *High-volume fly ash system: Concrete solution for sustainable development*. Materials Journal, 2000.**97**(1): p. 41–48.

26. Zhu, D.Y., M.Z. Rong, and M.Q. Zhang, *Self-healing polymeric materials based on microencapsulated healing agents: From design to preparation*. Progress in Polymer Science, 2015.**49**: p. 175–220.

27. He, Z., et al., *Facile and cost-effective synthesis of isocyanate microcapsules via poly-vinyl alcohol-mediated interfacial polymerization and their application in self-healing materials*. Composites Science and Technology, 2017.**138**: p. 15–23.

28. Otsuki, N., et al., *Influences of bending crack and water-cement ratio on chloride-induced corrosion of main reinforcing bars and stirrups*. Materials Journal, 2000.**97**(4): p. 454–464.

29. Kessler, M., *Self-healing: A new paradigm in materials design*. Proceedings of the Institution of Mechanical Engineers, Part G: Journal of Aerospace Engineering, 2007.**221**(4): p. 479–495.

30. Trask, R., H.R. Williams, and I. Bond, *Self-healing polymer composites: Mimicking nature to enhance performance*. Bioinspiration & Biomimetics, 2007.**2**(1): p. P1.

31. White, S.R., et al., *Autonomic healing of polymer composites*. Nature, 2001.**409**(6822): p. 794–797.

32. Li, H., et al., *Preparation and application of polysulfone microcapsules containing tung oil in self-healing and self-lubricating epoxy coating*. Colloids and Surfaces A: Physico-chemical and Engineering Aspects, 2017.**518**: p. 181–187.

33. Ghosh, S.K., *Self-healing materials: Fundamentals, design strategies, and applications*. 2009: Wiley Online Library, Germany, p. 1–307.

34. Therriault, D., S.R. White, and J.A. Lewis, *Chaotic mixing in three-dimensional micro-vascular networks fabricated by direct-write assembly*. Nature Materials, 2003.**2**(4): p. 265–271.

35. Therriault, D., et al., *Fugitive inks for direct-write assembly of three-dimensional micro-vascular networks*. Advanced Materials, 2005.**17**(4): p. 395–399.

3 Self-Healing Cementitious Materials

3.1 INTRODUCTION

The phenomenon of self-healing in concrete has been known for many years. It has been observed that some cracks in old concrete structures are lined with white crystalline material, suggesting the ability of concrete to self-seal the cracks with chemical products, perhaps with the aid of rainwater and carbon dioxide in air. Later, a number of researchers [1–4] in the study of water flow through cracked concrete under a hydraulic gradient noted a gradual reduction of permeability over time, again suggesting the ability of the cracked concrete to self-seal itself and slow the rate of water flow.

In recent years, there is increasing interest in the phenomenon of mechanical property recovery in self-healed concrete materials. For example, the resonance frequency of an ultra-high-performance concrete damaged by freeze/thaw action, and the stiffness of pre-cracked specimens, were demonstrated to recover after water immersion. In another investigation, the recovery of flexural strength was observed in pre-cracked concrete beams subjected to compressive loading at early age [5–6].

In these studies, self-healing was associated with continued hydration of cement within the cracks. As in previous permeability studies, the width of the concrete cracks, found to be critical for self-healing to take place, was artificially limited using feedback-controlled equipment and/or by the application of a compressive load to close the preformed crack. These experiments confirm that self-healing in the mechanical sense can be attained in concrete materials [6–7].

Deliberate engineering of self-healing in concrete was stimulated by the pioneering research of White et al., who investigated self-healing of polymeric material using encapsulated chemicals. A number of experiments were conducted on methods of encapsulation, sensing, and actuation to release the encapsulated chemicals into concrete cracks. For example, Ghosh et al. [8] demonstrated that air curing polymers released into a crack could lead to a recovery of the composite elastic modulus.

3.2 SELF-HEALING PROCESSES IN CONCRETE

Concrete is the most widely used man-made building material on the planet, and cement is used to make approximately 2.5 tons (more than 1 cubic meter) of concrete per person per year [9]. Concrete structures have been built since the discovery of Portland cement (PC) in the midst of the nineteenth century. The reaction of PC with water results in hydration products, which glue the reacting cement particles together to form a hardened cement paste. When cement and water are mixed with sand, the

DOI: 10.1201/9781003195764-3

resulting product is called mortar. If the mixture also contains coarse aggregate, the resulting product is called concrete. It is a quasi-brittle material, strong in compression but relatively week in tension. The compressive strength of traditional concrete varies between 20 and 60 MPa [10]. By using a low water–cement ratio, improved particle packing, and special additives, high strength concrete can be produced with strength values up to 150–200 MPa. Cement-based composites have even been produced with compressive strengths up to 800 MPa.

Concrete elements loaded in bending or in tension easily crack. For this reason, reinforcement is installed. Passive reinforcement is activated as soon as the concrete cracks. The formation of cracks is considered an inherent feature of reinforced concrete. It must be emphasized that in reinforced concrete structures, cracks as such are not considered as damage or failure, and cracking as such does not indicate a safety problem. The crack width, however, should not exceed a prescribed crack width limit. Cracks that are too wide may reduce the capacity of the concrete to protect the reinforcing steel against corrosion. Corrosion of reinforcing steel is the major reason for premature failure of concrete structures. Apart from these macro-cracks, very fine cracks—that is, micro-cracks—may occur within the matrix due to restraint of shrinkage deformations of the cement paste. Micro-cracks are an almost unavoidable feature of ordinary concrete. If micro-cracks form a continuous network of cracks, they may substantially contribute to the permeability of the concrete, thereby reducing the concrete's resistance against ingress of aggressive substances [11].

Even though cracks can be judged as an inherent feature of reinforced concrete and the existence of cracks does not necessarily indicate a safety problem, cracks are generally considered undesirable for several reasons [12]. The presence of cracks may reduce the durability of concrete structures. In cases where structures have to fulfill a retaining function, cracks may jeopardize the tightness of the structure. Completely tight concrete may be required in case the structure has to protect the environment against radiation from radioactive materials or radioactive waste.

Cracks may also be undesirable for aesthetic reasons. Not only cracks, but also the inherent porous structure of concrete can be a point of concern. If pores are connected and form a continuous network, harmful substances may penetrate the concrete and may chemically or physically attack the concrete or the embedded steel. The performance of structures with elapse of time is often presented with graphs [13]. After some time, gradual degradation occurs until the moment that first repair is urgently needed. The durability of concrete repairs is often a point of concern. Very often a second repair is necessary only 10–15 years later. Spending more money initially in order to ensure higher quality often pays off. The maintenance-free period will be longer and the first major repair work can often be postponed for many years. Apart from saving direct costs for maintenance and repair, the savings due to reduction of the indirect costs are generally most welcomed by the owner.

The experience is that an initial higher quality of the material results in postponement of repair and, consequently, a reduction of the costs for maintenance and repair. This raises a logical question concerning the optimum balance between increasing the initial costs and the cost savings for maintenance and repair. The extreme case would be that no costs for maintenance and repair have to be considered at all because the material is able to repair itself schematically, showing the performance of a structure made with self-repairing material. On the occurrence of a small crack or the start of any physical or chemical degradation process, the material gradually starts to repair itself and the structure will regain its original level of performance or a level close to that illustrates the anticipated costs for such a material. In this figure, inflation and interest are not considered. The initial costs will be substantially higher than that of a structure made with traditional concrete mixtures. The absence of maintenance and repair costs, however, could finally result in a financially positive situation for the owner.

An enhanced service life of concrete structures will reduce the demand for new structures. This, in turn, results in the use of less raw materials and associated reductions in pollution, energy consumption, and CO_2 production. Statistics convincingly show the enormous amounts of money spent by society owing to the lack of quality and durability of concrete structures. The cost for reconstruction of bridges in the United States has been estimated between \$20 and \$200 billion [14]. The average annual maintenance cost for bridges in that country is estimated at \$5.2 billion. Comprehensive life cycle analyses indicate that the indirect costs due to traffic jams and associated lost productivity are more than 10 times the direct cost of maintenance and repair [15]. In The Netherlands, one-third of the annual budget for large civil engineering works is spent on inspection, monitoring, maintenance, upgrading, and repair. In the United Kingdom, repair and maintenance costs account for over 45% of the annual expenditure on construction [16–17]. According to the Department for Environment, Food and Rural Affairs (DEFRA), about half of CO_2 emissions in the United Kingdom come from buildings, of which the actual structure is a large contributor. The production of 1 t of PC, for example, produces approximately 1 t of CO_2 when the emissions due to calcination and fuel combustion required to power the kiln are combined [9]. Considering that about 2.35×10^9 t of cement are used annually worldwide, the CO_2 emissions associated with the production of cement are very significant, and are estimated to be in the order of 5–7% of the total CO_2 production in the world. Given the rapid growth of China's and India's economies, which are the two largest consumers of cement, this figure is expected to increase if the technology to produce cement remains unchanged. Enhancing the longevity of our built infrastructure will undoubtedly reduce the impact of mankind's activities on the stability of the biosphere. Attempts to justify fundamental and risky research by promising the solutions of the serious environmental problems faced by society today may sound a bit modish. This, however, should not keep us from starting this research. Not starting this research will worsen the situation. Moreover, a direct benefit for many parties involved in the building industry and for society as a whole is conceivable if it were possible to improve the quality and service life of concrete structures. Historically, the concrete in these structures has been designed to meet predefined specifications at the start of its life. Longevity of the structure is then monitored through maintenance programs. More recently, however, material scientists have begun to adopt a change in philosophy whereby adaptability or "self-healing" of the material over time is explicitly considered. The inspiration for this change has often come from nature through biomimicry. In concrete, early research in this self-healing area has focused on both the natural ability of hydrates to heal cracks over time, and artificial means of crack repair from adhesive reservoirs embedded within the matrix. This chapter focuses on work that has been completed recently at Delft and Cardiff universities in both "natural" (autogenic) and "artificial" (auto-nomic) healing of concrete. Before this, a brief overview of the current state of the art in this area is given, in addition to an explanation of some commonly used definitions.

3.3 EXPANSIVE AGENTS AND MINERAL ADMIXTURES

In a study by Kishi et al. [18], it was found that cementitious materials like Al_2O_3–Fe_2O_3-tri (AFt), Al_2O_3–Fe_2O_3-mono (AFm), and calcium carbonate ($CaCO_3$) were created in cracked concrete and $Ca(OH)_2$ crystal air voids. The working assumption argued for the leaching out of such hydration yields and renewed crystallization in the flow water via fractures. In line with this, a range of repair agents—including expansive agents, geomaterials, and chemical mixtures, together with their blends—were used to assess the performance of the concrete in terms of its ability to repair itself [19–20]. In addition, comparative analysis was undertaken between the reference specimen and the specimen with 10% cement content replaced with expansive materials consisting of C_4A_3S, $CaSO_4$, and CaO. Results showed that the presence of the expansive agents in the concrete beams nearly enabled the repair of an early 0.22 mm crack after more than 30 days with the detection of rehydration yield between the cracks. In contrast, for the standard concrete structures, the partial repair of the cracks occurred during an identical interval of time. Hence, in comparison to the standard concrete, a higher efficiency was demonstrated by the expansive agent re-crystallization in the air voids for the self-repair [18]. Qureshi et al. [21] provided the evidence that the concrete mixtures containing the expansive minerals had a greater capability of repairing themselves. Representing a marker of the state of the cement mixture at a given age, the hydration degree enabled the quantitative estimation of the cement mixture performance in terms of the self-repair.

It was demonstrated that the geopolymer was formed due to the addition of the geomaterial with a content of 71.3% SiO_2 and 15.4% Al_2O_3 to the expansive material through a separate polymerization of the aluminate and silicate complexes [20]. The presence of the alkali metals caused the dissolution of the polymerized aluminate and silicate complexes at the alkaline pH. As uncovered by an extensive examination, the geopolymer gel particles were less than 2 μm in size, and a large number of hydro-garnet or Aft phases were produced by the cracked interfacial phases associated with the original ruptured zone (Figure 3.1). In comparison to the hydro-garnet phase, the dense phase contained most of the altered geopolymer gel as revealed by the EDX spectra. According to the additional analysis of the chemical additives, an improvement was achieved by the supplementing regular concrete enclosing the $NaHCO_3$, Na_2CO_3, and Li_2CO_3. Such composition triggered the cementitious re-crystallization and concrete particle precipitation [20]. The conclusion reached was that the crack self-repair could be greatly enhanced through the addition of suf-ficient quantities of the carbonates and expansive agents.

3.4 HOLLOW FIBERS

Self-repair is made possible by the hollow fibers, as illustrated in Figure 3.2. These fibers use the voids constituting a composite network matrix for the storage of certain

FIGURE 3.1 Microstructure of results between self-healing area and original area [22].

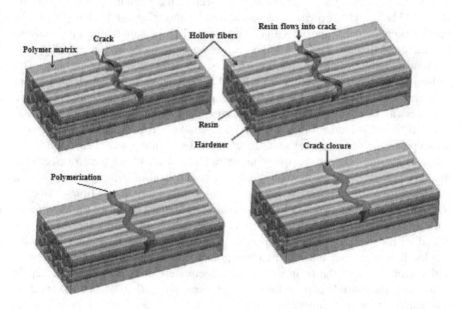

FIGURE 3.2 Schematic representation of self-healing processes using hollow fibers [34].

functional constituents of the materials serving as repair agents [23–24]. When the extrinsic stresses or stimuli cause concrete structure deterioration and crack formation, the functional constituents are released from the voids in which they are stored to automatically heal the damage. Thus, these hollow fibers work akin to arteries in a living organism, making it possible for the materials in which they are embedded to repair themselves [25]. For example, the self-repair of polymeric composites is facilitated by the bulk polymers [26–28]. The technique of damage visual enhancement was devised by Pang and Bond [29] for the rapid and straightforward detection of the internal damage in the composite structures. Evidence has been provided that the progress of the healing process can be monitored via the fibers containing engineering healing materials and labeled with fluorescent dye.

The technology of crack repair in the cementitious materials is based on the same mechanism as the self-healing biological functions similar to the blood coagulation [30–31]. It involves the incorporation of the functional constituents in the delicate fibrous vessels permeating the structural network of the concrete. The occurrence of the deterioration triggers the rupture of the fibers and release of the functional repair materials that activate the self-healing. Empirical work has succeeded in making concrete less porous by incorporating the liquid methyl methacrylate (MMA), methacrylic acid methyl ester, and reactive resin into the hollow polypropylene fibers that were subsequently inserted into the concrete [31]. Other researchers investigated the discharge of the crack-bridging cementitious glue from the hollow glass pipettes within the concrete instantly after the flexural test. Compared to the concrete structures without such glue constituents, those with the glue were capable of carrying 20% heavier loads.

The buoyant process associated with self-repair has been discussed in various studies [32–33]. This process was elucidated by inserting the hollow fibers in a cementitious network matrix with one end linked to the self-repair mediator and the other end tied up. The concrete mixtures were produced and inserted into the glass tubes with 2 mm external diameter and 0.8 mm internal diameter [32]. The self-repair constituent included the diluted (27%) and non-diluted alkali-silicate mixtures and two components integrating epoxy resin of low viscosity. Next, the loading was performed until the crack mouth opening displacement (CMOD) reached to 0.03–2 mm following the load removal. To determine whether the self-repair capability was enhanced, the cracked specimens were subjected to the renewed curing. Strength recovery mean ratios of 1.1 and 1.5 were respectively displayed by the specimens with the repair constituent of dilute and non-dilute alkaline silica solution, unlike the specimens without repair constituent. In contrast, the strength recovery ratio did not show significant gains for the specimens with epoxy resin. In fact, it was around three times lower than the ratio associated with the direct mixing and manual injection of the resin into areas with cracks. It was argued that the two components had not been properly blended and stirred, causing the resin to harden inadequately and result in a low repair ratio. Another possible explanation was that the pipes still contained the residual epoxy, as one end was blocked.

According to Joseph et al. [33], apart from a few negligible differences, the testing protocol was identical. The conduit of repair mediator (ethyl cyanoacrylate) was represented by bent plastic tubes with 4 mm external diameter and 3 mm internal diameter. The conclusion reached was that the self-repair can effectively be accomplished based on the external provision of such a repair constituent. Considerable improvement was noted in the post-crack stiffness, peak load, and ductility following the damage repair. As suggested by the observations made during and after assessment, the capillary suction and gravitational effects made the ethyl cyanoacrylate a suitable adhesive agent capable of permeating an extensive area of the surfaces with the cracks.

3.5 BACTERIA AS SELF-HEALING AGENT

Biological mechanisms for repair involving the introduction of the bacteria into the concrete have been recently suggested [4, 35–38]. During the middle of the 1990s, a sustainable method for healing the cracks in concrete was proposed by Gollapudi et al. [39] that involved the introduction of ureolytic bacteria to speed up $CaCO_3$ precipitation in the concrete micro-crack zones. Several parameters were employed to describe this process, including the amount of dissolved inorganic carbon, the pH of the material, the levels of calcium ions, and nucleation site accessibility. The walls of the bacterial cells represented the nucleation sites, while the other parameters were regulated by the bacterial metabolism [35]. Tittelboom et al. [36] used the bacteria in the concrete to generate an enzyme called the urease that could catalyze urea $(CO[NH_2]_2)$ into the ammonium ions (NH^{4+}) and carbonate radicals (CO_3^{-2}). In the chemical reactions, 1 mol of urea underwent the intracellular hydrolyses to 1 mol of carbonate and 1 mol of ammonia following Path I. Then, carbamate was hydrolyzed spontaneously to form one extra mole of ammonia and carbonic acid via path II. These products later formed 1 mol of bicarbonate (HCO_{-3}) and 2 moles of ammonium (NH^{+4}) and hydroxide (OH^{-1}) ions (Path III and IV). Path IV and V was responsible for the enhancement of pH, drifting the bicarbonate equilibrium to form carbonate ions.

$$CO(NH_2)_2 + H_2O \rightarrow NH_2COOH + NH_3 \qquad \text{(Path I)}$$
$$NH_2COOH + H_2O \rightarrow NH_3 + H_2CO_3 \qquad \text{(Path II)}$$
$$H_2CO_3 + H_2O \rightarrow HCO^{-3} + H^{+1} \qquad \text{(Path III)}$$
$$2NH_3 + 2H_2O \rightarrow 2NH^{+4} + 2OH^{-1} \qquad \text{(Path IV)}$$
$$HCO^{-3} + H^{+1} + 2NH^{+4} + 2OH^{-1} \rightarrow CO_3^{-2} + 2NH^{+4} + 2H_2O \qquad \text{(Path V)}$$

The walls of the bacterial cells had a negative charge, so cations from the surrounding environment can be accepted by the bacteria, with the deposition of Ca^{+2} on the cell wall surface. The reaction between the Ca^{+2} and CO_3^{-2} resulted in the precipitation of $CaCO_3$ on the surface of the cell wall, thus providing the active nucleation sites. The repair of the surface cracks was made possible through this method of the bacterial-based localized $CaCO_3$ precipitation as illustrated in Figure 3.3.

$$Ca^{+2} + Cell \rightarrow Cell\text{-}Ca^{+2} \qquad \text{(Path VI)}$$
$$Cell\text{-}Ca^{+2} + CO_{3-}^2 \rightarrow Cell\text{-}CaCO_3 \qquad \text{(Path VII)}$$

3.6 MICROENCAPSULATION

Inspired by the natural phenomena, several encapsulation materials have been made with diverse sizes, from macroscale to nanoscale. At the macroscopic level, the natural encapsulation in its most basic form is embodied by the bird eggs or seeds, whereas at the macroscopic level, the natural encapsulation is embodied by an egg or seed cell [40–42]. The starting point of the microencapsulation development was

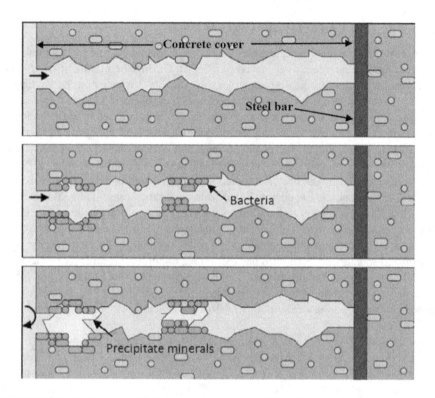

FIGURE 3.3   Typical crack-healing processes via immobilized bacteria in concretes [4].

the creation of the dye-containing capsules. Numerous novel technologies have been introduced in different fields of applications [43]. The microencapsulation is not a separate component, but can be understood as insertion of solid granules of the order of micrometers, liquid drops, or gases into the inert shell, affording protection against the activity of the external agents [44–45]. This was the basis of the micro-encapsulation with self-repair capability [46]. The mechanism of the self-repair is illustrated in Figure 3.4. The embedded microcapsules break when the cracks are formed, causing the release of the repair agent into the crack surfaces through capillary action, followed by the interaction between the repair agent and incorporated catalyst, thereby triggering the polymerization and sealing of the cracks. Figure 3.4b elucidates the microcapsule rupturing mechanism.

In a study conducted by Nishiwaki [47], the epoxy resin was used as a repair material packed into the microcapsules with the incorporation of the urea-formaldehyde formalin shell of size 20–70 μm. Microcapsules were used, together with the acrylic resin. The results indicated that the self-repair was successfully achieved by the microcapsules containing sodium silicate [2]. The concrete stacking was initially performed nearly to the breaking point prior to the removal of the load and subsequent curing for seven days. In comparison to the reference specimen that exhibited just

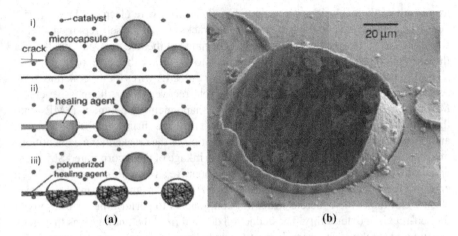

FIGURE 3.4 (a) The microcapsulation approach and (b) typical Field Emission Scanning Electron Microscope (FESEM) micrograph of a cracked microcapsule [44].

10% recovery, the specimen with 2% sodium silicate microencapsulation regained its strength to a proportion of up to 26%. It was concluded that the upper strength recovery ratio was possible by increasing the concentration of the repair material. Meanwhile, Du et al. [48] prepared the single-component microcapsules with the toluene-di-isocyanate (TDI) and paraffin serving as the repair agent and the shell, respectively. The findings revealed that the encapsulation of the TDI within the paraffin shell was successful and a better self-repair capability was demonstrated by mortars containing the microcapsules.

3.7 SHAPE MEMORY MATERIALS AS SELF-HEALER

The integration of the functional materials such as shape memory alloys (SMAs) or shape memory polymers into cementitious concrete structures to promote self-repair has been advocated in a number of studies [49–52]. The underlying principle is that the crack formation triggers the controlled shrinkage of these materials, producing a contraction serving to seal the cracks. The shape memory effect was discovered by Song et al. and reported that the gold-cadmium (Au-Cd) alloy exhibits a reversible phase change [52]. Since then, several SMAs with exceptional thermo-mechanical and thermo-electrical properties have been developed [53]. For example, nitinol is a highly elastic alloy that demonstrates the shape memory effect and capability of reversing to its pre-established form upon the heat exposure [54]. Its super-elasticity permits it to withstand a major inelastic deformation and returns it to its original form once the load is removed. Song and Mo [55] produced an intelligent reinforced concrete (IRC) by employing the SMA wires. More specifically, the stranded martensite wires of the SMAs were used to accomplish the post-tension effects in the IRC. The strain distribution in the concrete was obtained by tracking changes in the electrical resistance of the SMA wires. This allowed detection of the cracks forming as a result of the explosions or earthquakes. The SMA wire electrical heating triggered the contraction and thus alleviating the cracks so that the self-repair capability was effective for managing macro-cracks. The name of the IRC derived from the fact that the concrete structure is sufficiently smart to recognize the self-repair.

Sakai et al. [49] studied the super-elasticity of the SMA wires for the concrete beam self-repair and acknowledged almost a complete recovery after the massive crack. Meanwhile, Jefferson et al. [51] integrated the shape memory polymers (SMPs) into the cementitious materials and showed that the early-age shrinkage, thermal effects, and/or mechanical loading can cause the formation of the cracks in the cementitious matrix. Exposure to the heat can activate the shrinkage of the incorporated SMA tendons, which in turn generates quantifiable compressive stresses throughout the closed crack surfaces, thus promoting crack healing. This kind of mechanism of the crack closure can make the concrete structural components to perform better in terms of the self-repair and durability. The deduction derived from the findings was that crack closure and weak pre-stressing in post-tensioned mortar beams can be successfully addressed based on the incorporation of the parallel polymer tendons subjected to the shrinking. Numerous screening tests indicated that the tendons of highest efficiency are the polyethylene terephthalate (PET) shrinktite, exhibiting a shrinking potential of around 34 MPa in a controlled environment with heating up to 90°C followed by the cooling down to the ambient temperature. An increase of about 25% in the mortar strength was estimated to be promoted by heating plus additional curing.

3.8   COATING

The supplementary functionalities created by the latest material technology innovations brought the concept of "intelligent material" that refers to the ability to effectively respond to the extrinsic stimuli (e.g. temperature, light, humidity). This has led to the successful development and testing of the advanced construction materials such as coatings for the concrete with the self-repair capability and particular durability qualities [56–57]. For reinforced concrete, such coatings are devised to promote steel bar self-repair and minimize the deterioration due to corrosion. The newest research initiative of the self-repairing coatings has the potential to make a notable contribution to the efforts to combat the contemporary infrastructure degradation. Unlike the standard anti-corrosion coatings in which the efficiency is compromised by the slightest coating damage, the self-repairing coatings are capable of recovery from the damage. Thus, the efficiency of the self-repairing coatings remains unaffected [58], and therefore, this capability is likely to extend the use life of the steel rebar structures considerably [59]. Chen et al. [60] first explored the use of the self-repairing coatings for steel rebar. The epoxy coatings that were usually applied to the rebar structures can be substituted with the self-repairing coatings, especially in the northeast regions to protect the rebar against high levels of the corrosion.

Identification of the strategies for preventing the micro-crack formation in the concrete structures (e.g. roads, bridges, etc.) has been the focus of ample research, but no definitive conclusions have been reached so far [61]. Penetration of concrete by water, de-icing salt, and air is facilitated by cracks. Sub-zero temperatures cause the expansion of frozen water within the cracks, which thus become larger in size and cause the concrete to deteriorate faster upon the exposure to the road salt. However, the self-repair coatings for concrete protection have received far less research attention than the self-repair coatings with the anti-corrosion action for metal protection. Several studies were conducted on the self-repairing coatings that trigger the self-repair in response to the extrinsic crack formation or damage. Such coatings frequently incorporate the micro containers that break readily in the presence of the disruptive agents. The healing agents within these containers can extend the coatings lifespan by sealing the existing cracks. Meanwhile, the containers themselves can take various forms including the polyurethane microcapsules and microfilament tubes, but they rarely affect the coating mechanical properties. Various promising findings in this field warranted additional exploration for the practical uses [62–63].

Over the years, diverse organic and inorganic materials have been employed as the healing agents to achieve high-performance self-repairing. Rebar protection against corrosive agents (e.g. water, salts) commonly relies on epoxy coatings. Studies have also been dedicated to the creation of the polymer coatings capable of activating crack repair in response to the external stimuli (e.g. heat, pH alterations). The heat-responsive coatings became successful, and few of them can maintain their mechanical properties even after repeated heat cycles [64–65]. It has been reported that the polyelectrolyte nanocontainer coatings can respond to the changes in the pH within seconds [66]. These types of coatings have great potential for the further applications, especially due to their distinct mechanical properties that can completely be restored. Particular attention has been directed toward drying oils like tung oil and linseed oil, owing to their excellent repair capabilities and encapsulation [67]. Upon the exposure to air, the tung oil undergoes the polymerization to a coating characterized by toughness, glossiness, and imperviousness [68–69]. Due to such properties, drying oils are widely incorporated into the paints, varnishes, and printing inks. Samadzadeh et al. [70] accomplished the first encapsulation of tung oil. Assessment of the microcapsules' pull-off strength revealed that the urea-formaldehyde microcapsules adhered to the epoxy matrix more effectively than the industrial standards. In addition, the evaluation of the lifespan by immersing the damaged specimens into the sodium chloride solutions yielded a positive outcome. Compared to the epoxy coatings, a nine-fold extension of the use life was achieved by tung oil microcapsules following the damage.

3.9 ENGINEERED CEMENTITIOUS COMPOSITE

The ultra-ductile fiber-reinforced cementitious composite also called engineered cementitious composite (ECC) is a special type of concrete that was introduced at the beginning of the 1990s. ECC was continually refined over past two decades [71]. It is highly ductile (3–7%) and displays a tight crack size and a relatively reduced amount of fibers that does not exceed 2% by volume [72]. The distinguishing mechanical quality of the ECC is the metal-like feature. Furthermore, the ECC can withstand heavy loading following crack formation in the context of auxiliary distortions. The self-healing notion of the dry related to the bleeding was investigated by Li et al. [73] regarding the release of chemicals capable of sealing tensile cracks with ulterior air curing. In this manner, the composites without cracks can recover their mechanical properties. However, the self-repair process was displayed low efficiency in case of standard concrete, cement, or fiber-reinforced concrete because the tensile crack size is challenging to control in such materials. A decrease in the tensile load can promote the relentless multiplication of the local breaks within the crack width, leading to the rapid depletion of the repair agents. Hence, it is necessary to reduce the tensile crack width to within tens of micrometers for achieving a successful self-repair. The alternative is to change the mechanical properties of the composites with the use of glass pipes of extremely large size. This was highlighted by several other studies that drew the attention to the significance of the crack width [74].

The notion of self-repair is feasible for the ECC due to the major property of tight crack size control of this material, as explored by two distinct empirical studies [75]. The first study involved the use of scanning electron microscopy (SEM) for in situ testing of the ECC with one empty glass fiber without the healing chemicals under the condition of the applied load, while the second study involved the measurement of the flexural strength of the ECC incorporated glass fibers with ethyl cyanoacrylate as the sealing agent. To assess the efficiency of the sealing mediator for the repair purposes following the occurrence of the deterioration in the load cycles, both studies were conducted under the MTS load-frames. The SEM images showed the sensing and actuation processes, while the flexural stiffness recovery was indicative of the effect of regeneration. It was admitted that prior to implementation into practice, additional problems must be addressed.

The self-repair capability of the cementitious materials and the use of the external chemicals as repair glue in concrete were investigated by Li et al. [76], who analyzed the concrete matrix and its interaction with the exposed surroundings. The cracks were induced into the ECC before exposing it to a range of environmental conditions including water penetration and submersion, wetting and drying cycles, and chloride ions attack. The results revealed almost full recovery of the mechanical and transport properties, especially for the ECC pre-loaded with the tensile strain below 1%. The self-repair was promoted by the minute crack size, the low ratio of water to binder, and the abundant fly ash (FA) content via hydration and pozzolanic mechanisms.

Waste matter and/or by-products were employed by Zhou et al. [77] for local production of ECC materials. Both slag and limestone powders were used to design a number of mixtures, which were then subjected to analysis that involved measurement of tensile strain (2–3%) and crack stiffness. Results indicated that unlike the mixtures studied by Li et al. [76], the mixtures considered by Zhou et al. [77] had a higher concentration of blast furnace slag (BFS) and lithium slag (LS) instead of FA as well as a higher water–binder ratio (0.45–0.60). It was inferred that a significantly lower amount of unhydrated cementitious materials with curing lasting longer than 28 days was used than in the study by Li et al. [73]. Furthermore, compared to the ECC materials with rich FA content and low ratio of water to binder, the ECC materials with high BFS and lime putty (LP) content and relatively high water–binder ratio display similar self-repair capability. This conclusion was derived from the tight crack width, with the ECC self-repair depending on the availability of unhydrated cement and additional complementary products (e.g. BFS). In fact, the self-repair was promoted by a low water-cementitious material ratio and high proportion of cementitious specimen. Moreover, crack size was highlighted as important for hydration-reliant self-repair, as crack sealing could be achieved with minimal use of healing agent and crack bridging from both sides was facilitated.

Through the release of the healing agents, the presence of the microencapsulated modules generally promoted improvement of ECC micro-crack behavior and likelihood of crack formation. Consequently, the processes of sensing and actuation were made effective via the microencapsulation. As previously mentioned, a considerable importance is attached to the ECC tight crack size, as it minimizes the amount of healing agents needed to seal cracks and makes it easier for these agents to bridge cracks from both sides. In short, the self-repair capability of the ECC is greater compared to the standard cements because of its higher proportion of the cementitious materials and lower ratio of water to binder.

3.10 NANOMATERIALS-BASED SELF-HEALING CONCRETE

In the concrete industry, the use of self-repairing materials is still a relatively recent innovation. These materials refer to the materials that contain cement and are capable of autonomous recovery from deterioration caused by various factors. Simultaneously, considerable interest is currently being raised by the possibility of producing the sustainable concrete using nanomaterials. Thus, the concrete can be made more durable and sustainable by integrating the self-repair and nanomaterial technologies [78]. In the context of self-repair, nanomaterials have been employed primarily to mitigate the steel bar corrosion in the reinforced concrete. For instance, Koleva [78] indicated that the use of nanomaterials with customized qualities, such as the core-shell polymer vesicles or micelles, can help reinforced concrete to perform better. However, the use of nanomaterials in concrete with self-repair capability has not been extensively studied.

Qian et al. [79] investigated the curing under air, carbon dioxide, and water in both wet and dry states, as well as the impact of nanoclay with water employed as a hydration-related inner water furnishing agent on micro-cracks. The results revealed that the addition of nanoclay and more suitable amounts of cementitious materials into the mixtures could significantly enhance their repair capability. In addition, every air-cured mixture displayed the satisfactory repair as attested by the absence of the final crack formation at the new site. Meanwhile, De [38] discovered that the addition of the superabsorbent polymer capsules with water as an inner pool for supplementary hydration afforded the ECC a greater self-repair capacity. Furthermore, various repair products were identified over the cracked facades, but there were no obvious repaired cracks. It was concluded that the self-repair mechanism does not significantly affect the cracks. Access to sufficient water or moisture was also stressed as being important, as it not only served as the reactant for the supplementary hydration, but also facilitated the transport of ions.

3.11 SELF-HEALING IN FIBER-REINFORCED CONCRETE

Concrete composite is made of the binders, fine and coarse aggregates, and short length discontinuous fibers. These fibers can significantly enhance the impact resistance, ductility, and energy absorption of the concrete, in addition to higher values of splitting tensile and flexural strengths dispersed in the concrete mixture [80–81]. Many types of short fibers—including the metallic, polymeric, natural, and carbon— are used to reinforce the concrete for achieving enhanced properties. Earlier studies [82–83] on the fiber-reinforced cementitious composites showed an appreciable inhibition of the crack propagation, thus contributing to easy crack healing in the concrete. It was concluded that the inclusion of the fibers in the concrete matrix can enhance their engineering properties. In [83], authors evaluated the feasibility of the fibers-reinforced concrete as immobilizers wherein the self-healing agents such as bacteria were used. Rauf et al. [83] showed that the natural fibers (coir, flax, and jute) can be used to carry bacterial spores for self-healing of the concrete. For this purpose, the calcite precipitation bacteria—namely, the *Bacillus subtilis* KCTC-3135^T, *Bacillus cohnii* NCCP-666, and *Bacillus sphaericus* NCCP-313—were included in the concrete matrix, along with the calcium lactate pentahydrate and urea as the organic nutrients. It was found that the natural fibers are capable of substantial immobilization of the bacterial spores. In addition, the flax fibers provided better protection to the bacteria with improved crack-healing, and regained the compressive strength.

Figure 3.5 shows the microstructure of three selected concrete specimens incorporated with the bacteria and natural fibers (*B. subtilis* with jute, *B. cohnii* with flax, and *B. sphaericus* with coir fibers). As a carrier, the flax fibers provided a better protection to the bacteria, indicating an efficient crack-healing and pore-filling ability. Zhu et al. [84] developed some sustainable ECCs by incorporating the limestone calcined clay cement and polypropylene fibers. The results displayed that the proposed concrete can attain an efficient recovery of the composite tensile ductility and ultimate tensile strength through the self-healing mechanism. It was established that the inclusion of the fibers in the self-healing concrete matrix can affect positively the mechanical properties and self-healing efficiency of the studied concretes.

3.12 SUMMARY

In recent times, the production of sustainable concretes via self-healing technology became useful in the construction industries worldwide. Exponential increase in the usage of the OPC caused severe environmental damages. The immense benefits and usefulness of self-healing concrete technologies were demonstrated in terms of their sustainability, energy-saving traits, and environmental affability. The foremost challenges, current progress, and future trends of the smart technology related to self-healing concrete's potential were emphasized. An all-inclusive overview of the appropriate literature on the smart materials–based self-healing concretes allowed us to draw the following conclusions:

i. The self-healing concretes are characterized through several significant traits such as low pollution levels, cheap, eco-friendly, and elevated durability performance in harsh environmental conditions. These properties make these concretes the effective sustainable materials in the construction industries.

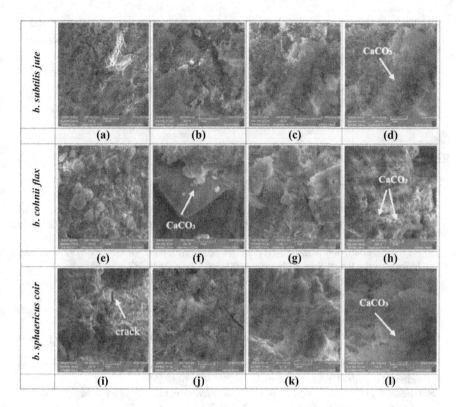

FIGURE 3.5 The SEM image of the healed specimens prepared with different types of fibers [84].

ii. The internal encapsulation and hollow fiber-activated self-healing strategies are efficient for the repair of multiple damages. However, these two strategies encounter some complexity in the casting and have negative impact on the mechanical properties of the proposed concretes.

iii. The inclusion of the expansive agents and mineral admixtures in the concrete demonstrated superior efficiency in the self-healing process. However, it is not very effective in the presence of multiple damages.

iv. The design of the nanomaterials-based self-healing concretes with improved performances and endurance are useful for several applications, thanks to the advancement of the nanoscience and nanotechnology.

v. The environmental pollution can considerably be reduced by implementing the high strength and durable cementitious composites fabricated using diverse nanoparticles, carbon nanotubes, and nanofibers.

vi. In the domain of the building and construction, the production of the materials via the nanotechnology route is going to play a vital role toward achieving sustainable development in the near future.

vii. The use of smart materials in concrete is advantageous in terms of the improved engineering properties of the cementitious materials, especially for the generation of self-healing and sustainable concretes.

viii. This comprehensive review is believed to provide taxonomy to navigate and underscore research progress toward the smart materials–based self-healing concrete technology.

REFERENCES

1. Behfarnia, K., *Studying the effect of freeze and thaw cycles on bond strength of concrete repair materials.* Asian Journal of Civil Engineering, 2010.**11**(2): p. 165–172.
2. Huseien, G.F., K.W. Shah, and A.R.M. Sam, *Sustainability of nanomaterials based self-healing concrete: An all-inclusive insight.* Journal of Building Engineering, 2019.**23**: p. 155–171.
3. Hayes, S., et al., *Self-healing of damage in fibre-reinforced polymer-matrix composites.* Journal of the Royal Society Interface, 2007.**4**(13): p. 381–387.
4. Jonkers, H.M., et al., *Application of bacteria as self-healing agent for the development of sustainable concrete.* Ecological Engineering, 2010.**36**(2): p. 230–235.
5. Jonkers, H.M., Self healing concrete: A biological approach. In *Self healing materials.* 2007: Springer, Netherlands. p. 195–204.
6. Dick, J., et al., *Bio-deposition of a calcium carbonate layer on degraded limestone by Bacillus species.* Biodegradation, 2006.**17**(4): p. 357–367.
7. Huseien, G.F., et al., *Performance of epoxy resin polymer as self-healing cementitious materials agent in mortar.* Materials, 2021.**14**(5): p. 1255.
8. Ghosh, S., et al., *Microbial activity on the microstructure of bacteria modified mortar.* Cement and Concrete Composites, 2009.**31**(2): p. 93–98.
9. Van Oss, H.G., *Background facts and issues concerning cement and cement data.* US Geological Survey Open File Report, 2005.**1152**: p. 44.
10. Shah, K.W., and G.F. Huseien, *Biomimetic self-healing cementitious construction materials for smart Buildings.* Biomimetics, 2020.**5**(4): p. 47.
11. Huseien, G.F., et al., *Geopolymer mortars as sustainable repair material: A comprehensive review.* Renewable and Sustainable Energy Reviews, 2017.**80**: p. 54–74.
12. Vijay, K., M. Murmu, and S.V. Deo, *Bacteria based self healing concrete—A review.* Construction and Building Materials, 2017.**152**: p. 1008–1014.
13. Van Breugel, K., *Is there a market for self-healing cement-based materials.* Proceedings of the first international conference on self-healing materials, 2007.
14. Yunovich, M., and N.G. Thompson, *Corrosion of highway bridges: Economic impact and control methodologies.* Concrete International, 2003.**25**(1): p. 52–57.
15. Freyermuth, C.L., *Life-cycle cost analysis for large bridges.* Concrete International, 2001.**23**(2): p. 89–95.
16. Schlangen, E., *Self-healing phenomena in cement-based materials.* RILEM. See www.rilem. net/tcDetails. php, 2005.
17. Schlangen, E., and C. Joseph, *Self-healing processes in concrete.* 2009: Self-Healing Materials: Fundamentals, Design Strategies, and Applications. Germany, Wiley-VCH. p. 1–307.
18. Kishi, T. *Self healing behaviour by cementitious recrystallization of cracked concrete incorporating expansive agent.* 1st International conference on self healing materials, Noordwijk, The Netherlands, 2007. p. 4.
19. Ahn, T.-H., and T. Kishi, *Crack self-healing behavior of cementitious composites incorporating various mineral admixtures.* Journal of Advanced Concrete Technology, 2010.**8**(2): p. 171–186.

20. Al-Ansari, M., et al., *Performance of modified self-healing concrete with calcium nitrate microencapsulation.* Construction and Building Materials, 2017.**149**: p. 525–534.
21. Qureshi, T., A. Kanellopoulos, and A. Al-Tabbaa, *Autogenous self-healing of cement with expansive minerals-II: Impact of age and the role of optimised expansive minerals in healing performance.* Construction and Building Materials, 2019.**194**: p. 266–275.
22. Tae-Ho, A., and T. Kishi, *Crack self-healing behavior of cementitious composites incorporating various mineral admixtures.* Journal of Advanced Concrete Technology, 2010.**8**(2): p. 171–186.
23. Kousourakis, A., and A. Mouritz, *The effect of self-healing hollow fibers on the mechanical properties of polymer composites.* Smart Materials and Structures, 2010.**19**(8): p. 085021.
24. Zhong, N., and W. Post, *Self-repair of structural and functional composites with intrinsically self-healing polymer matrices: A review.* Composites Part A: Applied Science and Manufacturing, 2015.**69**: p. 226–239.
25. Dry, C., *Procedures developed for self-repair of polymer matrix composite materials.* Composite Structures, 1996.**35**(3): p. 263–269.
26. Motuku, M., U. Vaidya, and G. Janowski, *Parametric studies on self-repairing approaches for resin infused composites subjected to low velocity impact.* Smart Materials and Structures, 1999.**8**(5): p. 623.
27. Bleay, S., et al., *A smart repair system for polymer matrix composites.* Composites Part A: Applied Science and Manufacturing, 2001.**32**(12): p. 1767–1776.
28. Pang, J., and I. Bond, *'Bleeding composites'—damage detection and self-repair using a biomimetic approach.* Composites Part A: Applied Science and Manufacturing, 2005.**36**(2): p. 183–188.
29. Pang, J.W., and I.P. Bond, *A hollow fibre reinforced polymer composite encompassing self-healing and enhanced damage visibility.* Composites Science and Technology, 2005.**65**(11–12): p. 1791–1799.
30. Dry, C., *Matrix cracking repair and filling using active and passive modes for smart timed release of chemicals from fibers into cement matrices.* Smart Materials and Structures, 1994.**3**(2): p. 118.
31. Dry, C., *Three designs for the internal release of sealants, adhesives, and waterproofing chemicals into concrete to reduce permeability.* Cement and Concrete Research, 2000.**30**(12): p. 1969–1977.
32. Mihashi, H., et al., *Fundamental study on development of intelligent concrete characterized by self-healing capability for strength.* Transactions of the Japan Concrete Institute, 2000.**22**: p. 441–450.
33. Joseph, C., A. Jefferson, and M. Cantoni, *Issues relating to the autonomic healing of cementitious materials.* 1st International conference on self-healing materials, 2007.
34. Ghosh, S.K., *Self-healing materials: Fundamentals, design strategies, and applications.* 2009: Wiley Online Library, Germany. p. 1–307.
35. Van Tittelboom, K., et al., *Use of bacteria to repair cracks in concrete.* Cement and Concrete Research, 2010.**40**(1): p. 157–166.
36. Luo, J., et al., *Interactions of fungi with concrete: Significant importance for bio-based self-healing concrete.* Construction and Building Materials, 2018.**164**: p. 275–285.
37. Algaifi, H.A., et al., *Numerical modeling for crack self-healing concrete by microbial calcium carbonate.* Construction and Building Materials, 2018.**189**: p. 816–824.
38. De Belie, N., et al., *A review of self-healing concrete for damage management of structures.* Advanced Materials Interfaces, 2018.**5**(17): p. 1800074.
39. Gollapudi, U., et al., *A new method for controlling leaching through permeable channels.* Chemosphere, 1995.**30**(4): p. 695–705.
40. Hemsley, A.R., and P.C. Griffiths, *Architecture in the microcosm: Biocolloids, self-assembly and pattern formation.* Philosophical Transactions of the Royal Society of London. Series A: Mathematical, Physical and Engineering Sciences, 2000.**358**(1766): p. 547–564.

41. Bekas, D., et al., *Self-healing materials: A review of advances in materials, evaluation, characterization and monitoring techniques.* Composites Part B: Engineering, 2016.**87**: p. 92–119.

42. He, J., and X. Shi, *Developing an abiotic capsule-based self-healing system for cementitious materials: The state of knowledge.* Construction and Building Materials, 2017.**156**: p. 1096–1113.

43. Wu, M., B. Johannesson, and M. Geiker, *A review: Self-healing in cementitious materials and engineered cementitious composite as a self-healing material.* Construction and Building Materials, 2012.**28**(1): p. 571–583.

44. White, S.R., et al., *Autonomic healing of polymer composites.* Nature, 2001.**409**(6822): p. 794–797.

45. Li, W., Z. Jiang, and Z. Yang, *Acoustic characterization of damage and healing of microencapsulation-based self-healing cement matrices.* Cement and Concrete Composites, 2017.**84**: p. 48–61.

46. Sun, D., et al., *Fatigue behavior of microcapsule-induced self-healing asphalt concrete.* Journal of Cleaner Production, 2018.**188**: p. 466–476.

47. Nishiwaki, T., *Fundamental study on development of intelligent concrete with self-healing capability.* Master's Thesis, 1997.

48. Du, W., et al., *Preparation and application of microcapsules containing toluene-di-isocyanate for self-healing of concrete.* Construction and Building Materials, 2019.**202**: p. 762–769.

49. Sakai, Y., et al., Experimental study on enhancement of self-restoration of concrete beams using SMA wire. In *Smart structures and materials 2003: Smart systems and nondestructive evaluation for civil infrastructures.* International Society for Optics and Photonics, CA, Japan, 2003.**5057**. p. 178–186.

50. El-Tawil, S., and J. Ortega-Rosales, *Prestressing concrete using shape memory alloy tendons.* Structural Journal, 2004.**101**(6): p. 846–851.

51. Jefferson, A., et al., *A new system for crack closure of cementitious materials using shrinkable polymers.* Cement and Concrete Research, 2010.**40**(5): p. 795–801.

52. Song, G., N. Ma, and H.-N. Li, *Applications of shape memory alloys in civil structures.* Engineering Structures, 2006.**28**(9): p. 1266–1274.

53. Seifried, F., et al., *Structure, morphology and selected mechanical properties of magnetron sputtered (Mo, Ta, Nb) thin films on NiTi shape memory alloys.* Surface and Coatings Technology, 2018.**347**: p. 379–389.

54. Shen, D., et al., *Development of shape memory polyurethane based sealant for concrete pavement.* Construction and Building Materials, 2018.**174**: p. 474–483.

55. Song, G., and Y. Mo, *Increasing concrete structural survivability using smart materials.* A proposal submitted to grants to enhance and advance research (GEAR), University of Houston, 2003.

56. Pittaluga, M., *The electrochromic wall.* Energy and Buildings, 2013.**66**: p. 49–56.

57. Gobakis, K., et al., *Development and analysis of advanced inorganic coatings for buildings and urban structures.* Energy and Buildings, 2015.**89**: p. 196–205.

58. Lau, K., and A.A. Sagüés, *Coating condition evaluation of epoxy coated rebar.* ECS Transactions, 2007.**3**(13): p. 81.

59. Shchukin, D.G., et al., *Active anticorrosion coatings with halloysite nanocontainers.* The Journal of Physical Chemistry C, 2008.**112**(4): p. 958–964.

60. Chen, Y., et al., *Self-healing coatings for steel-reinforced concrete.* ACS Sustainable Chemistry & Engineering, 2017.**5**(5): p. 3955–3962.

61. Samadi, M., et al., *Enhanced performance of nano-palm oil ash-based green mortar against sulphate environment.* Journal of Building Engineering, 2020.**32**: p. 101640.

62. Ariffin, N.F., et al., *Strength properties and molecular composition of epoxy-modified mortars.* Construction and Building Materials, 2015.**94**: p. 315–322.

63. Nesterova, T., K. Dam-Johansen, and S. Kiil, *Synthesis of durable microcapsules for self-healing anticorrosive coatings: A comparison of selected methods.* Progress in Organic Coatings, 2011.**70**(4): p. 342–352.

64. Yang, W.J., et al., *Antifouling and antibacterial hydrogel coatings with self-healing properties based on a dynamic disulfide exchange reaction.* Polymer Chemistry, 2015.**6**(39): p. 7027–7035.

65. Luo, X., and P.T. Mather, *Shape memory assisted self-healing coating.* ACS Macro Letters, 2013.**2**(2): p. 152–156.

66. Kötteritzsch, J., et al., *One-component intrinsic self-healing coatings based on reversible crosslinking by Diels—Alder Cycloadditions.* Macromolecular Chemistry and Physics, 2013.**214**(14): p. 1636–1649.

67. Andreeva, D.V., et al., *Self-healing anticorrosion coatings based on pH-sensitive polyelectrolyte/inhibitor sandwichlike nanostructures.* Advanced Materials, 2008.**20**(14): p. 2789–2794.

68. Suryanarayana, C., K.C. Rao, and D. Kumar, *Preparation and characterization of microcapsules containing linseed oil and its use in self-healing coatings.* Progress in Organic Coatings, 2008.**63**(1): p. 72–78.

69. Jadhav, R.S., D.G. Hundiwale, and P.P. Mahulikar, *Synthesis and characterization of phenol—formaldehyde microcapsules containing linseed oil and its use in epoxy for self-healing and anticorrosive coating.* Journal of Applied Polymer Science, 2011.**119**(5): p. 2911–2916.

70. Samadzadeh, M., et al., *Tung oil: An autonomous repairing agent for self-healing epoxy coatings.* Progress in Organic Coatings, 2011.**70**(4): p. 383–387.

71. Li, V.C., *From micromechanics to structural Engineering-the design of cementitous composites for civil engineering applications.* Japan Society of Civil Engineering, 1993.**10**(2): p. 37–48.

72. Zhou, J., et al., *Development of engineered cementitious composites with limestone powder and blast furnace slag.* Materials and Structures, 2010.**43**(6): p. 803–814.

73. Li, V.C., Y.M. Lim, and Y.-W. Chan, *Feasibility study of a passive smart self-healing cementitious composite.* Composites Part B: Engineering, 1998.**29**(6): p. 819–827.

74. Reinhardt, H.-W., and M. Jooss, *Permeability and self-healing of cracked concrete as a function of temperature and crack width.* Cement and Concrete Research, 2003.**33**(7): p. 981–985.

75. Qian, S., et al., Influence of microfiber additive effect on the self-healing behavior of engineered cementitious composites. In *Sustainable construction materials.* 2013: ASCE, United States. p. 203–214.

76. Li, V.C., and E.-H. Yang, Self healing in concrete materials. In *Self healing materials.* 2007: Springer, Germany. p. 161–193.

77. Zhou, J., et al., *Developing engineered cementitious composite with local materials.* International conference on microstructure related durability of cementitious composites, Nanjing, China, 2008.

78. Koleva, D., *Nano-materials with tailored properties for self healing of corrosion damages in reinforced concrete, IOP self healing materials.* 2008: SenterNovem, The Netherlands.

79. Qian, S., J. Zhou, and E. Schlangen, *Influence of curing condition and precracking time on the self-healing behavior of engineered cementitious composites.* Cement and Concrete Composites, 2010.**32**(9): p. 686–693.

80. Mohammadhosseini, H., M.M. Tahir, and A.R.M. Sam, *The feasibility of improving impact resistance and strength properties of sustainable concrete composites by adding waste metalized plastic fibers.* Construction and Building Materials, 2018.**169**: p. 223–236.

81. Mohammadhosseini, H., A.A. Awal, and J.B.M. Yatim, *The impact resistance and mechanical properties of concrete reinforced with waste polypropylene carpet fibers.* Construction and Building Materials, 2017.**143**: p. 147–157.

82. Homma, D., H. Mihashi, and T. Nishiwaki, *Self-healing capability of fiber reinforced cementitious composites.* Journal of Advanced Concrete Technology, 2009.**7**(2): p. 217–228.

83. Rauf, M., et al., *Comparative performance of different bacteria immobilized in natural fibers for self-healing in concrete.* Construction and Building Materials, 2020.**258**: p. 119578.

84. Zhu, H., et al., *Mechanical and self-healing behavior of low carbon engineered cementitious composites reinforced with PP-fibers.* Construction and Building Materials, 2020.**259**: p. 119805.

4 Self-Healing Measurement Methods

4.1 CONCRETE DURABILITY

The most prevalently utilized construction material is concrete, as its compressive strength, affordability, availability of its raw materials, and durability are unparalleled. Nonetheless, time wears away at concrete within infrastructure and general structures, causing degradation slowly yet persistently. The integral structure and efficacy of concrete degrades due to the permeation of water. The ingress of water, undesired acidic gases, and fluids containing dissolved particles occurs in part through crack formations at the macro and micro scales [1]. As a result, aggressive substances and the mentioned materials pass through these permeable sites. Hence, the durability and reinforcement of concrete is negatively affected, which in turn implies that the performance of concrete in the long term is predicated by the environment in which it resides [2–3]. Reinforced steel bars are oxidized and weakened by the permeation of water and the durability of exposed concrete infrastructure is similarly negatively affected [4–5]. Accessibility to certain cracks are impossible, as they cannot be seen. The quantity and size of crack formations proliferate as the materials undergo permeation, contraction, and expansion. Consequently, an onus is increasingly placed upon maintenance and inspection methods for the mentioned materials and infrastructure. The substantial costs associated with constant maintenance, which are particularly exorbitant for large-scale infrastructure, demonstrate the challenges of adopting frequent inspection and maintenance methods. Maintenance and reparations can also be challenging due to damaged site in the infrastructure being difficult to access. Resultantly, academics in the field increasingly focus upon the marginal capital and labor needs associated with self-healing concrete that can autonomously repair hazardous cracks. Thus, the mentioned marginal labor and capital needs of self-healing concrete make investigating the capabilities of its self-healing properties via diverse approaches an interesting prospect. This chapter utilizes diverse approaches to evaluate the efficacy of the self-healing properties of this kind of concrete. The efficacy of the self-healing properties is determined by noting whether the cement base material reflects its original form and functionality following the autonomous repair of a crack formation.

Recovering and healing entirely or in part a material's functionality following degradation makes up the definition of self-healing materials [6]. Similarly, autonomously repairing and identifying impairment is another definition of self-healing materials [7]. Thus, manual intervention is unnecessary within the reparation process [8]. Properties such as the durability of concrete, and more, are enhanced by the relatively novel evolution of biotechnology and nanotechnology.

DOI: 10.1201/9781003195764-4

4.2   SELF-HEALING TECHNIQUES

Self-healing concrete is capable of repairing crack formations at the micro scale while also restoring the mechanical properties of the cement base materials. The generation of self-healing concrete, in line with sustainable development, has involved multiple academic investigations into replicating the natural healing process of bones via bacteria alongside chemical and autogenous self-healing [9–10]. Self-healing concrete has been improved through the integration of microorganisms, hollow fibers, microencapsulation, polymers, mineral admixtures, and other expansive materials [9].

Additional hydration of unreacted cement, the expansion of hydrated cementitious matrix, obstructing cracks with impurities in the water, and calcium carbonate formation are methods that can lead to the organic occurrence of autonomous healing [11]. Moreover, self-healing properties in concrete were generated via the utilization of geomaterials, polymers, and chemical admixtures [12]. Another form of generating self-healing properties in concrete was achieved through calcium carbonate precipitating microorganisms [13–14]. The ability of concrete to autonomously repair itself is enabled and enhanced by the amalgamation of mineral admixtures and additions alongside cementitious materials in materials with expansive properties. Nevertheless, cement in close proximity with water must be safeguarded to avoid the cement reacting with the water and prematurely expanding.

Crack formations at the micro scale can be filled in via calcium and magnesium silicates being dissolved in the bulk water that makes the concrete [15]. However, these methods do not repair cracks in their entirety; rather, the plugging effect diminishes the size of the cracks. Through tensile pre-loading, high-strength concrete is damaged—and these simulated crack formations subsequently demonstrate self-healing properties when exposed to water and laid bare to the environment [16]. Cementitious composite incorporated synthetic fibers in the mentioned scenario. As a result, the complete healing of the simulated cracks occurred [17]. Furthermore, crack formations at the micro scale were autonomously repaired more efficiently via the use of supplementary cementing materials [11]. Improved self-healing of crack formations in the concrete while also diminishing consumption of cement are among the benefits of these supplementary materials [18]. Additionally, high-performance fiber can be used to reinforce such cementitious composite [19].

Upon undergoing distinct curing regimes, the aforementioned approaches were outstripped with regard to self-healing efficacy. Oil water, sea water, and water submersion made up these distinct curing regimes. The effective plugging of cracks only occurred when smaller than 50 μm in size, thus demonstrating the limits of this form of autonomous healing. Rather than using still water, a constant water leakage was used by researchers to heal crack formations in concrete, demonstrating an enhanced autonomous healing effect [20]. Shrinkable polymers have been utilized to induce post-tensioning of concrete, which results in improved self-healing efficacy [21]. Research has focused upon engineered cementitious composite due to this, discovering that such engineered composites enhance autonomous healing performance, as well as being an autonomous healing material with great potential [22]. Other academics found that when ECC is combined with limestone powder, the restoration of functionality is almost entirely achieved [23]. Further research has been undertaken on carbonated steel slag and whether it effectively acts as a self-healing agent in concrete. A size of 20 lm wide and 5 mm long was discovered to be the maximum size for this method. The width of filled in cracks are evaluated via various diverse observation methods, such as X-ray computed tomography, light microscope, ultrasound, and camera photographs [24–25].

Extending the service life and autonomous healing of concrete structures via the integration of polymers into the concrete mix has been the aim of some academics [26]. Encouraging findings were achieved through this method. Nonetheless, the polymer type determined the efficacy of self-healing [27]. Additionally, the efficacy of autonomous healing was affected by cement-to-water ratio, kind of cement, and polymer dose. Enhanced autonomous healing was also induced via the amalgamation of separate materials with the polymer [26]. A complete restoration in compressive strength can possibly be achieved by certain polymer-based self-healing agents [28]. Almost two-thirds of a crack formation's width can be plugged through the autonomous healing catalyzed by a polymer, as per another academic work [29].

An elevated ultrasonic pulse velocity, as well as the control specimen being surpassed by 16% regarding flexural strength restoration, were affirmed in academic works [30–31]. Additionally, diverse methods for autonomously healing concrete have been compared [32]. The method of utilizing a superabsorbent polymer was found to be surpassed by enacting encapsulation technique for autonomously healing concrete, as per the mentioned comparative work. As the autonomous healing is induced without the use of water, the encapsulation techniques can be implemented in a greater range of contexts.

It is possible that organic characteristic in building materials can be simulated via the use of nanomaterials that embody cutting edge technology in the field of nanotechnology [33–34]. Sought-after characteristics are generated in a novel material that is produced via the combination of nanoparticles into the concrete mix [33]. Very high levels of chemical reactivity are produced in the novel material as nanoparticles have a high ratio of surface area to volume. As a result, the generation of a novel material with ideal properties is possible. Thus, the production of new and sustainable cementitious composites with greater performance levels are enabled by nanotechnology. The evolution of self-healing concrete could advance through the use of functionalized silica nanoparticles, a material with great potential which is highlighted in an academic work [35]. Hence, further investigation is required within this domain. Other research examined how the durability characteristics of cement-based materials were affected by different nanoparticles [36]. In addition, concrete's waterproof performance was improved via the utilization of nanocomposite, as per a recent study [1].

Another area of research that has been backed by various academic involves examining the efficiency of bacteria in concrete as an autonomously healing agent [37–38]. Crack formations in the concrete were all efficiently plugged in through the use of mixed and isolated bacterial cultures [39]. The metabolic activity of the bacteria provoked the precipitation of calcium carbonate, which in turn filled in the cracks [40]. Autonomous healing can be catalyzed by the injection of a bacterial culture into the surface of the concrete [13]. Additionally, in a parking garage, crack formations in the concrete had bacterial cultures sprayed directly onto them [41]. Resultantly, autonomous healing led to a substantial decrease in water permeability. Nonetheless, there was no substantial restoration of the compressive strength. Because of this, concrete and cement mortar's durability and compressive strength were elevated through the use of microorganisms in many studies [42]. The use of admixtures in concrete in combination bacteria led to additional enhancements to the compressive strength [43]. Such additions led to a 36% increase in the compressive strength [44]. To this day, concrete durability and strength are enhanced via the incorporation of bacteria in multiple ongoing studies.

A distinct method to the injection and spray technique was enacted by a separate study that instead placed bacteria directly into the concrete mix. This led to the ureolytic activity of the bacteria provoking microbial precipitation, which in turn filled in the crack formations. While this method surpasses alternative techniques, the bacteria's lifespan was diminished due to the concrete matrix being an unforgiving environment for the bacteria. Hence, as time passed, the efficacy of autonomous healing diminished [45]. The presence of a nucleation site, the concentration of the calcium ion, pH, and the concentration of dissolved inorganic carbon are among the elements that lead to microbial precipitation. The added bacteria is substantially affected by the high alkalinity of the concrete mix, which makes it a harsh habitat. Resultantly, the lifespan can only be enhanced via safeguarding the bacteria [46]. The encapsulation of the bacteria is one method to accomplish the safeguarding of the bacteria. In such a case, the autonomous healing properties of the concrete mix are improved. Through the use of polyurethane, a bacterial culture was immobilized in the concrete matrix which, in turn, led to the gradual reduction of enzymatic activity and the prolongation of the bacteria's lifespan [47]. Nonetheless, for a prolonged length of time, the stabilization of enzymatic activity is also provoked.

Immobilizing the bacteria was also found to cause a 60% restoration in compressive strength, as per an academic work [48]. Similarly, the addition of ureolytic bacteria to the concrete matrix was carried out by other researchers [49]. The gradual filling of crack formations occurred via this method, as precipitation of calcium carbonate within the crack formation was provoked. Interestingly, the more time passed, the stronger was the performance of the immobilized bacteria [50]. Moreover, the concrete was combined with a hydrogel containing encapsulated bacterial spores in other studies, which led to a crack measuring 0.5 mm wide being plugged in its entirety while a near 70% decrease in water absorption was also observed [51]. Academic works have also demonstrated how a crack width of almost 1000 lm was plugged in its entirety via the improved performance of an immobilized bacteria within a microcapsule [52]. In the present, graphite nanoplatelets have been used to encapsulate bacteria [53]. Specimens or crack formations pre-cracked at seven and three days were plugged in entirely when the width of the crack was 0.81 mm as a healing agent is secreted from the capsule after being stimulated by the crack formations. Throughout the concrete matrix, an even distribution of nanoplatelets can be achieved, which is one of the prime benefits of using nanoplatelets. Such benefits also include filling crack formations at the nanoscale, as well as enhancing compressive strength by almost 10%.

Overall, efficacy of a healing agent, processing survivability, and thermal stability are needs that must be met by encapsulated materials aiming to be utilized as healing agents. In the case of bacteria with granules within them, forming a coating agent from geopolymer is outlined to have great potential by certain academics [54]. Its implementation as an autonomous healing agent within concrete can achieve stellar results. At the same time, research was done on the potential autonomous healing agent of a double-walled sodium silicate microcapsule [55]. A remarkable level of performance was enacted by the mentioned microcapsule, efficiently activating autonomous healing with the concrete.

The profundity of filled in crack formations served as the basis for an assessment of autonomous healing in additional studies [55]. Filled-in crack formations 27.2 mm and 32 mm deep were outlined in these papers. The extended performance and lifespan of bacteria when utilizing the encapsulation technique was found to improve the efficacy of self-healing in concrete beyond other techniques and approaches. Additionally, this approach was capable of plugging in larger crack formations in their entirety. Alternatively, evolving autonomously healing concrete via the addition of chemical and biological agents revealed itself to be a technique with great potential [56]. This is exemplified by the complete filling in of a crack formation measuring 0.22 mm wide.

4.3 SELF-HEALING MEASUREMENT METHODS

The width of plugged-in crack formations were principally assessed through the use of a microscope by the vast bulk of researchers [57–58]. Techniques enacted included X-ray computed tomography, photographs from cameras with a very high pixel capacity, and digital images [59]. The occurrence of natural phenomenon plugged in a maximum crack formation size of 60 lm, as illustrated in Table 4.1. Meanwhile, polymer was utilized to plug in a 138 lm wide crack formation. Moreover, crack formations measure 200 lm wide were filled in via the utilization of supplementary cementing material within the concrete mix. Upon preparing the concrete, enacting the technique of encapsulating microorganisms led to large crack formations measuring 970 lm wide, which reflects progressive enhancements.

The profundity of the crack formation serving as a basis for assessing autonomous healing was a method enacted by not many academics, as per the literature [60]. A singly study used the length of a crack formation to assess self-healing. The encapsulation technique was utilized to plug in the maximum crack formation depth of 32 mm, while microorganisms were used to plug in a 27.2 mm deep crack formation,

TABLE 4.1
Approach and Measured Variables of Self-Healing

No.	Approach	Width of the Crack
1	Natural	Healing of crack below 60 μm wide
2	Supplementary cementing material	Crack width below 200 μm wide
3	Polymer and epoxy	Crack width up to 138 μm wide
4	Bacteria	Maximum crack width of 0.970 mm

TABLE 4.2

Techniques of Self-Healing Materials

No.	Approach	Depth of the Crack
1	Microencapsulation	Maximum depth of 32 mm
2	Bacteria	Maximum depth of 27.2 mm

as illustrated in Table 4.2. Thus, as illustrated in Table 4.1 and Table 4.2, the best approach was revealed to be the encapsulation technique, given that the maximum plugged-in crack formation depth and width were 32 mm and 0.97 mm, respectively. A maximum crack formation length of 5 mm was filled in by only one other study [61].

4.4 EFFICACY OF AUTONOMOUS HEALING TECHNIQUES

The efficacy of autonomous healing techniques have been assessed via a variety of techniques and evaluations within the relevant literature. Various aspects affect the autonomous healing process in crack formations. The age of crack formations, the curing regime, and the width of crack formations are typical aspects affecting said process [62]. The findings from a variety of academics, techniques to assess independent variables, and the independent and control variables are illustrated in Table 4.3. Improved autonomous healing with both immobilized and free organisms was found to occur due to the increased precipitation of calcium carbonate that stems from an elevated concentration of bacteria in the concrete mix, as shown in Table 4.3. Similarly, augmenting the polymer dosage also augments self-healing capacity. Autonomous healing was only improved due to the magnesium within bulk water, within the natural self-healing process, where the calcium in the bulk water had no positive effect. The enactment of diverse curing regimes upon engineered cementations concrete was found to improve the efficacy of water curing, as demonstrated in Table 4.3.

Additionally, upon biological and chemical agents being added to the concrete mix one at a time, the latter agent was found to enhance autonomous healing more than the former. The sole utilization of a microcapsule was also found to be less effective than bio encapsulation. Autonomous healing was affected less significantly by moisture and air curing techniques, thereby highlighting the more effective water curing method, as deduced by the illustrated comparison of diverse curing regimes in Table 4.3. As a crack formation grows increasingly wide, the efficacy of autonomous healing diminishes, as per Table 4.3, which also demonstrates that the implementation of nano-based materials in this field was carried out by a single academic work [53]. The mentioned study revealed that a crack formation 0.81 mm wide was filled in entirely, while an almost 10% increase in compressive strength was observed.

4.5 STRUCTURE TESTS FOR EVALUATING SELF-HEALING EFFICIENCY

The hardened form of concrete and its quality criteria are determined via structure tests at nanoscale, microscale, and macroscale. Nanostructure, microstructure, and

TABLE 4.3

Control Variables and Methods of Self-Healing Materials.

No.	Control Variables	Methods	Results
1	Immobilized organism	Optical density	More ureolytic activity with immobilized samples for longer period
2	Laboratory condition	Optical density	More self-healing at optimum concentration
3	Microfibers	Weight present	Plug cracks of up to 138 lm completely
4	Medium cultures	Liner method	Mixed culture of bacterial strains was able to plug cracks; however, compressive strength was not significantly regained
5	Ambient conditions	Particle size analysis	More self-healing with microcapsule of 230 μm size

TABLE 4.4

Structure Tests for Evaluating Self-Healing Efficiency

No.	Test	Dependent Variable
1	Strength properties	Compressive, flexural, and splitting tensile strength
2	Permeability	Ultrasonic pulse velocity, chloride penetration, gas permeability
3	Scanning electronic microscopy (SEM)	Fabric of nanostructure of concrete
4	Nanostructure measurement	Nanomechanical value

macrostructure assessments formed part of these tests. Table 4.4 illustrates how these assessments measure the efficacy of autonomous healing. Hence, the cement-based material self-healed sufficiently after cracking to meet the required quality criteria and restore its functionality. Macrostructure level assessments were undertaken by most academics to ascertain the autonomous healing effectiveness, as per the literature. The dependability of the findings in some works were augmented by undertaking microstructure assessments. Nonetheless, assessments at the nanostructure scale were rarely undertaken in the literature.

4.5.1 Macrostructure Tests

Multiple assessments at the macro level were carried out to ascertain self-healing effectiveness. The extent to which flexural and compressive strength, as well as other mechanical properties, recover is tested by these assessments, as illustrated in Table 4.4. Toughness and split tensile assessments form part of these tests [63–65]. Various academics mention that ultrasonic pulse velocity and water permeability assessments also form part of these tests [66–67]. Encapsulated tetraethyl orthosilicate, colloidal silica, sodium silicate, and their respective effectiveness in autonomously healing are evaluated by some researched via gas permeability and sorptivity assessments [68]. These assessments revealed an 18% decrease in sorptivity and a 69% decrease in gas permeability. The healed material's performance is measured via a stiffness assessment [69]. The control specimen was found to have 50% less flexural strength than the healed concrete [30]. In the crack formations, the white deposited material was assessed and recognized via thermo gravimetric analysis [26]. As demonstrated in Table 4.4, several academics undertook oxygen profile and chloride penetration assessments alongside pore size distribution and porosity assessments.

Following autonomous healing, up to 60% of the compressive strength can be recovered, as per the findings. Thus, the biological self-healing technique could be suitable for the durability criterion over a large stretch of time with greater critical effects, as per the findings illustrated in Table 4.4. Concrete infrastructure and its lifespan can be prolonged by such an approach. Nonetheless, some assessments are rarely undertaken, such as permeability, stress-strain relationship curve, bonding strength between concrete, and the deposited materials within crack formations. Hence, multiple techniques must be used to accurately assess the development of autonomously healing concrete. As a result, the suitability with the concrete composition and the bonding ability of deposited material in crack formations will be compared more effectively.

4.5.2 MICROSTRUCTURE TESTS

Overall, following the process of autonomous healing, the deposited material in concrete crack formations is analyzed and identified through the mentioned assessments, which serve to enhance the dependability of the findings. Examples of these assessments performed in the field include X-ray diffraction (XRD), field emission scanning electron microscopy (FESEM), and scanning electron microscopy (SEM). The deposited materials residing within crack formations are identified through the use of a SEM [70]. Distinct bacterial strains and their calcium carbonate precipitation, as well as polymerized and hydration products, make up these materials. Raman spectroscopy is frequently used in the present to evaluate autonomous healing performance [7]. The EDS, XRD, and SEM findings are presented in Table 4.4. The dependability of utilizing bacteria as an autonomous healing agent are optimized via this method [38]. Additionally, the test samples' crack formations were known to contain the deposition of such irregular crystals due to the findings from utilizing microstructures. Hence, while signal transmission rate of ultrasonic pulse velocity rises, acid ingress, chloride permeability, and water absorption fall significantly.

4.5.3 NANOSTRUCTURE TESTS

The dependability of the findings is optimized by undertaking assessments at the nano level. Nanostructure tests have been used to assess the efficacy of autonomous healing in recent academic works [71–72]. As Table 4.4 demonstrates, there is a strong bond between the deposited layer and the concrete matrix, as the average nanomechanical values at the outer precipitates are 20% less than in the transition zone. The bonding strength between the cement-based material, or the substrate, and the deposited materials at the interface can be evaluated by undertaking nano and macro level assessments, which in turn enhances this method's dependability.

4.6 COMPARISON OF ASSESSMENT TESTS

Multiple diverse techniques are used to carry out several assessments, as seen in Table 4.5. Measuring the efficacy of autonomous healing via the macro mechanical properties of a material is a method enacted by most academics. Moreover, such a method focusing upon macrostructure properties, on top of durability assessments, was deemed to be the optimal method by various academics as its findings are seemingly more dependable. Despite findings being less dependable, some academics carried out durability tests following the use of a microscope to view a crack formation's width. The sole use of a microscope by some academics embodied a method with lesser intensity. Findings that are more dependable are obtained via microscopic observation in tandem with microstructure assessment. To ascertain self-healing effectiveness, macro and microscopic mechanical tests were undertaken by a very small number of academics, as illustrated in Table 4.5. Microstructure and macrostructure assessments were carried out to determine autonomous healing efficacy, whereby the former were carried out in tandem with durability tests in some cases. The sole use of durability tests to ascertain autonomous healing efficacy was deemed to be one of the least dependable methods. Nano-level assessments were carried out to ascertain autonomous healing efficacy by two academics, as seen in Table 4.5. The macro mechanical test, the nanostructure and microstructure assessments, and the durability test were deemed to have never occurred at the same time, as per the comprehensive literature review. Increased dependability in determining autonomous healing efficacy can be deduced in future studies by forming an approach and method that integrates the mentioned diverse techniques.

TABLE 4.5
Scale Structural Tests Adopted for Evaluation of Self-Efficiency

No.	Scale	Tests
1	Macrostructure test	Ultrasonic pulse velocity, Compressive strength, flexural, toughness, stiffness, water permeability, chloride permeability
2	Microstructure test	Water absorption,
3	Microscopic test	Visual observation, flexural strength
4	Nanostructure tests	Flexural strength, ultrasonic pulse velocity, nanoscale mechanical measurement

4.7   SUMMARY

This chapter analyzed and compared several tests and methods that were developed to evaluate self-healing efficiency, reaching the following conclusions:

 i. Self-healing was measured in terms of decrease in the crack width, length, and depth via visual observation using microscope and digital imaging, camera photography, and X-ray computed tomography.
 ii. Recovery of strength properties was the common test adopted by majority of the authors to evaluate self-healing efficiency. Other studies have further conducted tests at macro and microstructure scales to maximize the reliability of the results. Moreover, very few studies have conducted tests at nanostructure scale.
 iii. There is a lack of availability of standard procedures or a main referenced procedure to measure rates of self-healing in cementitious materials.

REFERENCES

1. Muhammad, N.Z., et al., *Waterproof performance of concrete: A critical review on implemented approaches.* Construction and Building Materials, 2015.**101**: p. 80–90.
2. Basheer, L., J. Kropp, and D.J. Cleland, *Assessment of the durability of concrete from its permeation properties: a review.* Construction and Building Materials, 2001.**15**(2–3): p. 93–103.
3. Huseien, G.F., et al., *Evaluation of alkali-activated mortars containing high volume waste ceramic powder and fly ash replacing GBFS.* Construction and Building Materials, 2019.**210**: p. 78–92.
4. Aldea, C.-M., S. Shah, and A. Karr, *Permeability of cracked concrete.* Materials and Structures, 1999.**32**(5): p. 370–376.
5. Huseien, G.F., et al., *Geopolymer mortars as sustainable repair material: A comprehensive review.* Renewable and Sustainable Energy Reviews, 2017.**80**: p. 54–74.
6. Hager, M.D., et al., *Self-healing materials.* Advanced Materials, 2010.**22**(47): p. 5424–5430.
7. Bekas, D., et al., *Self-healing materials: A review of advances in materials, evaluation, characterization and monitoring techniques.* Composites Part B: Engineering, 2016.**87**: p. 92–119.
8. Guo, Y., et al., *Current progress on biological self-healing concrete.* Materials Research Innovations, 2015.**19**(sup8): p. S8–750–S8–753.
9. Wu, M., B. Johannesson, and M. Geiker, *A review: Self-healing in cementitious materials and engineered cementitious composite as a self-healing material.* Construction and Building Materials, 2012.**28**(1): p. 571–583.
10. Sangadji, S., and E. Schlangen, *Mimicking bone healing process to self repair concrete structure novel approach using porous network concrete.* Procedia Engineering, 2013.**54**: p. 315–326.
11. Talaiekhozani, A., and M.Z. Abd Majid, *A review of self-healing concrete research development.* Journal of Environmental Treatment Techniques, 2014.**2**(1): p. 1–11.
12. Vickers, N.J., *Animal communication: When I'm calling you, will you answer too?* Current Biology, 2017.**27**(14): p. R713–R715.
13. Sangadji, S., et al., *Injecting a liquid bacteria-based repair system to make porous network concrete healed.* ICSHM2013 4th international conference on self-healing materials, Ghent, Belgium, 2013, Magnel Laboratory for Concrete Research.

14. Schlangen, E., and S. Sangadji, *Addressing infrastructure durability and sustainability by self healing mechanisms-Recent advances in self healing concrete and asphalt*. Procedia Engineering, 2013.**54**: p. 39–57.

15. Parks, J., et al., *Effects of bulk water chemistry on autogenous healing of concrete*. Journal of materials in Civil Engineering, 2010.**22**(5): p. 515–524.

16. Yang, Y., E.-H. Yang, and V.C. Li, *Autogenous healing of engineered cementitious composites at early age*. Cement and Concrete Research, 2011.**41**(2): p. 176–183.

17. Nishiwaki, T., et al., *Self-healing capability of fiber-reinforced cementitious composites for recovery of watertightness and mechanical properties*. Materials, 2014.**7**(3): p. 2141–2154.

18. Huang, H., G. Ye, and D. Damidot, *Effect of blast furnace slag on self-healing of microcracks in cementitious materials*. Cement and Concrete Research, 2014.**60**: p. 68–82.

19. Ahn, T., D. Kim, and S. Kang, *Crack self-healing behavior of high performance fiber reinforced cement composites under various environmental conditions*. Earth and Space 2012: Engineering, Science, Construction, and Operations in Challenging Environments, 2012. p. 635–640.

20. Hosoda, A., et al., Self-healing properties with various crack widths under continuous water leakage. In *Concrete repair, rehabilitation and retrofitting II*. 2008: CRC Press, Florida, United States. p. 139–140.

21. Lv, Z., and D. Chen, *Overview of recent work on self-healing in cementitious materials*. Materiales de Construcción, 2014.**64**(316): p. e034–e034.

22. Yıldırım, G., et al., *A review of intrinsic self-healing capability of engineered cementitious composites: Recovery of transport and mechanical properties*. Construction and Building Materials, 2015.**101**: p. 10–21.

23. Siad, H., et al., *Influence of limestone powder on mechanical, physical and self-healing behavior of engineered cementitious composites*. Construction and Building Materials, 2015.**99**: p. 1–10.

24. In, C.-W., et al., *Monitoring and evaluation of self-healing in concrete using diffuse ultrasound*. NDT & E International, 2013.**57**: p. 36–44.

25. Wang, J., et al., *X-ray computed tomography proof of bacterial-based self-healing in concrete*. Cement and Concrete Composites, 2014.**53**: p. 289–304.

26. Snoeck, D., et al., *Self-healing cementitious materials by the combination of microfibers and superabsorbent polymers*. Journal of Intelligent Material Systems and Structures, 2014.**25**(1): p. 13–24.

27. Abd_Elmoaty, A.E.M., *Self-healing of polymer modified concrete*. Alexandria Engineering Journal, 2011.**50**(2): p. 171–178.

28. Van Tittelboom, K., and N. De Belie, *Self-healing in cementitious materials—A review*. Materials, 2013.**6**(6): p. 2182–2217.

29. Hazelwood, T., et al., *Numerical simulation of the long-term behaviour of a self-healing concrete beam vs standard reinforced concrete*. Engineering Structures, 2015.**102**: p. 176–188.

30. Ariffin, N.F., et al., *Mechanical properties and self-healing mechanism of epoxy mortar*. Jurnal Teknologi, 2015.**77**(12).

31. Sam, A.R.M., et al., *Performance of epoxy resin as self-healing agent*. Jurnal Teknologi, 2015.**77**(16).

32. Van Tittelboom, K., et al., *Comparison of different approaches for self-healing concrete in a large-scale lab test*. Construction and Building Materials, 2016.**107**: p. 125–137.

33. Sanchez, F., and K. Sobolev, *Nanotechnology in concrete—A review*. Construction and Building Materials, 2010.**24**(11): p. 2060–2071.

34. Sobolev, K., et al., *Application of nanomaterials in high-performance cement composites*. Proceedings of the ACI session on nanotechnology of concrete: Recent developments and future perspectives—2006, 2008. p. 93–120.

35. Perez, G., et al., *Synthesis and characterization of epoxy encapsulating silica microcapsules and amine functionalized silica nanoparticles for development of an innovative self-healing concrete.* Materials Chemistry and Physics, 2015.**165**: p. 39–48.
36. Morsy, M., S. Alsayed, and M. Aqel, *Hybrid effect of carbon nanotube and nano-clay on physico-mechanical properties of cement mortar.* Construction and Building Materials, 2011.**25**(1): p. 145–149.
37. De Muynck, W., N. De Belie, and W. Verstraete, *Microbial carbonate precipitation in construction materials: A review.* Ecological Engineering, 2010.**36**(2): p. 118–136.
38. Jonkers, H.M., and E. Schlangen, Development of a bacteria-based self healing concrete. In *Proceedings of the international FIB symposium.* 2008: Tailor made concrete structures, Taylor and Francis, London. p. 1–6.
39. Talaiekhozani, A., et al., *Application of proteus mirabilis and proteus vulgaris mixture to design self-healing concrete.* Desalination and Water Treatment, 2014.**52**(19–21): p. 3623–3630.
40. Jonkers, H.M., and E. Schlangen, A two component bacteria-based self-healing concrete. In *Concrete repair, rehabilitation and retrofitting II.* 2008: CRC Press, Florida, United States. p. 137–138.
41. Wiktor, V., and H. Jonkers, *Field performance of bacteria-based repair system: Pilot study in a parking garage.* Case Studies in Construction Materials, 2015.**2**: p. 11–17.
42. Ghosh, P., et al., *Use of microorganism to improve the strength of cement mortar.* Cement and Concrete Research, 2005.**35**(10): p. 1980–1983.
43. Achal, V., X. Pan, and N. Özyurt, *Improved strength and durability of fly ash-amended concrete by microbial calcite precipitation.* Ecological Engineering, 2011.**37**(4): p. 554–559.
44. Achal, V., A. Mukherjee, and M.S. Reddy, *Microbial concrete: Way to enhance the durability of building structures.* Journal of Materials in Civil Engineering, 2011.**23**(6): p. 730–734.
45. Jonkers, H.M., et al., *Application of bacteria as self-healing agent for the development of sustainable concrete.* Ecological Engineering, 2010.**36**(2): p. 230–235.
46. Van Tittelboom, K., et al., *Use of bacteria to repair cracks in concrete.* Cement and Concrete Research, 2010.**40**(1): p. 157–166.
47. Bang, S.S., J.K. Galinat, and V. Ramakrishnan, *Calcite precipitation induced by polyurethane-immobilized Bacillus pasteurii.* Enzyme and Microbial Technology, 2001.**28**(4–5): p. 404–409.
48. Wang, J., et al., *Use of silica gel or polyurethane immobilized bacteria for self-healing concrete.* Construction and Building Materials, 2012.**26**(1): p. 532–540.
49. Irwan, J., and N. Othman, An overview of bioconcrete for structural repair. In *Applied mechanics and materials.* 2013: Trans Tech Publ, Switzerland, **389**. p. 1–4.
50. Qian, C.X., et al., Self-healing and repairing concrete cracks based on bio-mineralization. In *Key engineering materials.* 2015: Trans Tech Publ, Switzerland, **629**. p. 494–503.
51. Wang, J., et al., *Application of hydrogel encapsulated carbonate precipitating bacteria for approaching a realistic self-healing in concrete.* Construction and Building Materials, 2014.**68**: p. 110–119.
52. Wang, J., et al., *Self-healing concrete by use of microencapsulated bacterial spores.* Cement and Concrete Research, 2014.**56**: p. 139–152.
53. Khaliq, W., and M.B. Ehsan, *Crack healing in concrete using various bio influenced self-healing techniques.* Construction and Building Materials, 2016.**102**: p. 349–357.
54. De Koster, S., et al., *Geopolymer coating of bacteria-containing granules for use in self-healing concrete.* Procedia Engineering, 2015.**102**: p. 475–484.
55. Mostavi, E., et al., *Evaluation of self-healing mechanisms in concrete with double-walled sodium silicate microcapsules.* Journal of Materials in Civil Engineering, 2015.**27**(12): p. 04015035.

56. Mors, R.M., and H.M. Jonkers, Bacteria-based self-healing concrete: An introduction. In *V international PhD student workshop on durability of reinforced concrete: From composition to service life design. VTT technology 65.* 2012: VTT Technical Research Center, Finland. p. 32–39.

57. Muhammad, N.Z., et al., *Tests and methods of evaluating the self-healing efficiency of concrete: A review.* Construction and Building Materials, 2016.**112**: p. 1123–1132.

58. Feiteira, J., E. Gruyaert, and N. De Belie, *Self-healing of dynamic concrete cracks using polymer precursors as encapsulated healing agents.* 2014: Concrete Solutions; Taylor & Francis Group, Ghent, Belgium. p. 65–69.

59. Liu, B., et al., *Trigger of self-healing process induced by EC encapsulated mineralization bacterium and healing efficiency in cement paste specimens.* 5th International Conference on Self-Healing Materials', Durham North Carolina, 2015.

60. Achal, V., A. Mukerjee, and M.S. Reddy, *Biogenic treatment improves the durability and remediates the cracks of concrete structures.* Construction and Building Materials, 2013.**48**: p. 1–5.

61. Pang, B., et al., *Autogenous and engineered healing mechanisms of carbonated steel slag aggregate in concrete.* Construction and Building Materials, 2016.**107**: p. 191–202.

62. Luo, M., C.-X. Qian, and R.-Y. Li, *Factors affecting crack repairing capacity of bacteria-based self-healing concrete.* Construction and Building Materials, 2015.**87**: p. 1–7.

63. Li, W., et al., *Self-healing efficiency of cementitious materials containing microcapsules filled with healing adhesive: Mechanical restoration and healing process monitored by water absorption.* PLoS One, 2013.**8**(11): p. e81616.

64. Cao, Q.Y., T.Y. Hao, and B. Su. Crack self-healing properties of concrete with adhesive. In *Advanced materials research.* 2014: Trans Tech Publ, Switzerland, **919**. p. 1880–1884.

65. Luhar, S., and S. Gourav, *A review paper on self healing concrete.* Journal of Civil Engineering Research, 2015.**5**(3): p. 53–58.

66. Sarkar, M., et al., *Autonomous bioremediation of a microbial protein (bioremediase) in Pozzolana cementitious composite.* Journal of Materials Science, 2014.**49**(13): p. 4461–4468.

67. Dong, B., et al., *A microcapsule technology based self-healing system for concrete structures.* Journal of Earthquake and Tsunami, 2013.**7**(03): p. 1350014.

68. Kanellopoulos, A., T. Qureshi, and A. Al-Tabbaa, *Glass encapsulated minerals for self-healing in cement based composites.* Construction and Building Materials, 2015.**98**: p. 780–791.

69. Granger, S., G. Pijaudier-Cabot, and A. Loukili. *Mechanical behavior of self-healed ultra high performance concrete: from experimental evidence to modeling.* The 6th international conference on fracture mechanics of concrete and concrete structures, Catalina, Italy, 2007.

70. Huang, H., and G. Ye, *Self-healing of cracks in cement paste affected by additional Ca^{2+} ions in the healing agent.* Journal of Intelligent Material Systems and Structures, 2015.**26**(3): p. 309–320.

71. Xu, J., and W. Yao, *Multiscale mechanical quantification of self-healing concrete incorporating non-ureolytic bacteria-based healing agent.* Cement and Concrete Research, 2014.**64**: p. 1–10.

72. Huseien, G.F., K.W. Shah, and A.R.M. Sam, *Sustainability of nanomaterials based self-healing concrete: An all-inclusive insight.* Journal of Building Engineering, 2019.**23**: p. 155–171.

5 Polymers-Based Self-Healing Cementitious Materials

5.1 INTRODUCTION

Fissures and cracks are the main concerns related to concrete due to its relatively low tensile strength. Cracks are detrimental not only in terms of facilitating corrosion, but also concerning the aesthetics because they make the porous structure in the concrete visible, expanding in size if no remedial action is taken. Hazardous chemicals can permeate concrete through the large cracks, leading to the chemical or physical deterioration of the concrete [1]. Several mechanical, chemical, and physical processes such as temperature gradients, shrinkage, external loading, and expansive reactions can create local stress and subsequently induce cracking [2–4]. Networks of continuous cracks significantly impair durability and permeability of concrete, providing an easy path for the entrance of gases and liquids that contain destructive substances that results in crack growth. Without proper and immediate treatment, such crack networks tend to spread and finally require costly treatment.

From a crack treatment perspective, conventional systems concentrate on the application of regular maintenance and scheming inspection. Nevertheless, in large-scale concrete structures, such conventional systems required a considerable amount of funds and labor. Besides, the treatment process somehow may be difficult or even impossible to implement where the affected structures are in service, i.e., large dams and tunnels. In these circumstances, self-healing concrete was proposed and has become increasingly attractive [5].

In concrete structures, self-healing of micro-cracks has already been introduced as a phenomenon that crack networks could be automatically repaired by rehydration of insufficiently hydrated or unhydrated cement particles in cracked areas. However, previous research indicated that the crack width that can be healed in such technique is limited to a maximum of 0.1 mm, which is not enough for practical application [6]. In addition to autogenous healing, cracks may also be automatically closed using a specific healing agent in the cement matrix [7–9].

Self-healing properties can be improved in concrete using different techniques including encapsulation of polymers, inclusion of fibers, and secondary hydration of unhydrated cement [10]. The existing self-healing agents, including cementitious materials, are classified broadly into four groups: intrinsic healing [11], microbial healing [12], capsule-based healing [13], and vascular healing [14]. Another alternative self-healing material that has attracted many researchers is bacterially produced calcium carbonate [4, 15]. Gollapudi et al. [16] first proposed this technique in the

DOI: 10.1201/9781003195764-5

mid-1990s to repair cracks using environmentally friendly processes. The technique involved incorporating ureolytic bacteria that facilitate enzymatic hydrolysis of pre-cipitation of calcium carbonate and urea in the micro-crack regions.

The epoxy resin's unique properties such as strong adhesion, high corrosion resistance, and chemical durability made it attractive in the construction sector to produce modified concretes. Also, epoxy resin is advantageous as the admixture in concrete because it can impart many emergent properties absent in the pure con-crete. However, to produce the epoxy resin–modified mortar, hardeners are essential [17–19]. According to Ariffin et al. [20–21], the inclusion of epoxy resin in concrete imparts hardening properties when alkaline substances are present. They developed the epoxy resin–modified mortars in the absence of any hardener. Ohama et al. [22] demonstrated that the epoxy resin–modified concretes and mortars in the absence of any hardener at (23 ± 3°C) have a lesser hardening rate and strength performance than those without epoxy resin. To overcome this drawback, the autoclave and steam curing techniques were followed to improve the early compressive strength (CS) developments of the epoxy resin–modified concretes [23–24]. Jo [25] observed that the characteristics of the epoxy resin–modified ordinary Portland cement (OPC)-based mortars in the absence of any hardener has relatively better properties than those with hardener. The observed increase in the strength performance was ascribed to the cross-linkages between the epoxy and the OPC network that could react with the $Ca(OH)_2$ without any hardener [26].

The reaction ability between the epoxy resin and hydroxyl ions from the OPC hydration $(Ca[OH]_2)$ to produce the hardened epoxy motivated the researchers to apply epoxy resin as a self-healing agent in concrete. It is established that the inclu-sion of the epoxy resin in the hydration process of the OPC matrix is responsi-ble for enhancing the mechanical properties of the product. In the OPC hydration process, the produced hydroxyl ions are essential for the self-healing mechanism and Calcium (Aluminum) Silicate Hydrate (C-(A)-S-H) gels formation. In the bacteria-based self-healing concretes, the bacterial strain reacts with the oxygen to produce the calcite, thus precipitating and healing the concrete's micro-cracks. In the epoxy resin–based self-healing concretes, similar mechanisms are involved, and also the epoxy resin can react with the hydroxyl groups, leading to healing of the concrete cracks.

In this chapter, the epoxy resin without any hardener—called diglycidyl ether of bisphenol A—was adopted as the self-healing agent to produce a sustainable mortar for several construction applications with long lifespan and high durability. The mixes were designed with various ratios of the epoxy resin to OPC to assess the efficacy of the epoxy as the self-healing agent. The modified mortars were subjected to dry and wet-dry conditions of curing to determine their improved properties such as flowability, CS development, porosity, degree of damage, and healing efficiency. The CS, ultrasonic pulse velocity (UPV), and SEM morpholo-gies were recorded to determine the feasibility of implementing the epoxy resin as the self-healing agent in mortar manufacturing. In addition, an optimized artificial neural network (ANN) combined with a firefly algorithm was developed to esti-mate the degree of damage and healing efficiency output parameters during the self-healing process.

5.2 MIX DESIGN AND SPECIMENS' PREPARATION

In this study, the OPC was collected from the Holcim Cement Manufacturing Company (Malaysia), following ASTM C150 standard [27] to prepare the mortar. The X-ray fluorescence (XRF) spectra were analyzed to determine the main chemical compositions of the OPC by weight percentage, including the CaO (63.1%), SiO_2 (20.1%), and Al_2O_3 (5.4%), whereby the total of loss on ignition (LOI) was 2.2% of the weight percentage. The mineral compounds of OPC, provided by the X-ray diffraction (XRD) pattern analyses, consist of dicalcium silicate (C_2S), tricalcium silicate (C_3S), tricalcium aluminate (C_3A), and calcium aluminoferrite (C_4AF) with the proportion of 9.3%, 74.9%, 6.7%, and 9.1%, respectively, as shown in Figure 5.1. The low amorphous part of OPC was shown between 2θ 25–55 degrees. The XRD pattern of OPC exhibits high pecks of C_2S and C_3S. These kind of elements will contribute to formulate the portlandite ($Ca[OH]_2$) during the hydration process. Silicate minerals in non-crystalline or an amorphous state are highly reactive with $Ca(OH)_2$ produced from the hydration of cement to form additional calcium silicate hydrate (C-S-H) gels.

To develop self-healing properties in the mix design, epoxy resin was used without any hardener for the easy reaction with the hydroxyl ions and hardening. Therefore, epoxy resin with high viscosity was selected, where it was obtained using a Brookfield digital viscometer, while the epoxy equivalent, molecular weight, and its flash point were provided by the manufacturing company. Table 5.1 summarizes the previously mentioned physical properties of epoxy resin.

The Fourier transform infrared (FTIR) reading of the epoxy resin and OPC were considered to identify and quantify their bonding structures, compositions, functional groups, and hydration properties. Figure 5.2 demonstrates the main FTIR

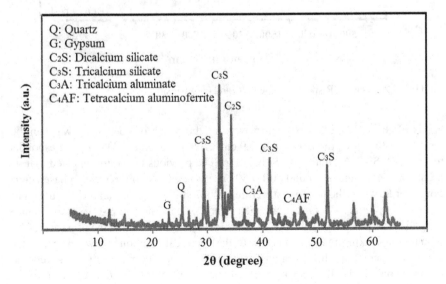

FIGURE 5.1 Cement XRD patterns.

TABLE 5.1

Physical Properties of Epoxy Resin

Epoxy equivalent	184
Molecular weight	380
Flash point (°C)	264
Viscosity (cPs, 20°C)	10000
Density (g/cm3, 20°C)	1.16

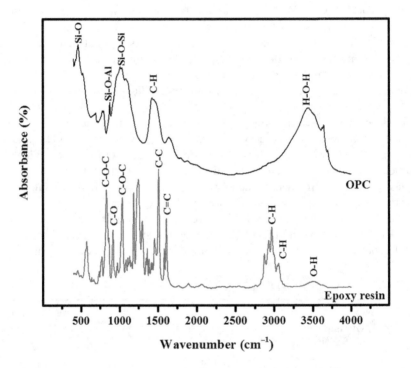

FIGURE 5.2   The FTIR spectra of the OPC and epoxy resin.

bands of the OPC and epoxy resin were in the range of 450–4000 wavenumber (cm⁻¹). For the OPC, the bands at 462.9 cm⁻¹, 824 cm⁻¹, and 1033 cm⁻¹ corresponded to the Si-O, Si-O-Al, and Si-O-Si bending. The previous literature reported that the broad band's existence around 500–650 cm⁻¹ was related to the glassy phase (short-ranged ordering of the network structures) of the silicates [28]. Meanwhile, the bands at 1426.7 cm⁻¹ and 3452 cm⁻¹ were due to the C-H bending and H-O-H (water bands) stretching, respectively. The epoxy resin's FTIR bands indicated the presence of different cross-linking mechanisms due to the chemical reactions between one or two kinds of monomers with different functional groups. The reactivity of the epoxies became completely different because of the alteration of the molecular structure. González et al. [29] showed the existence of highly reactive oxirane groups originated

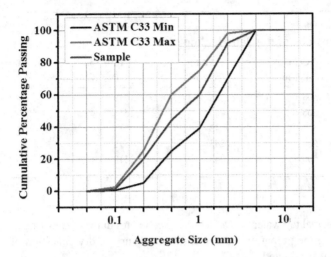

FIGURE 5.3   Particle size analysis of the fine aggregates.

from the linkages between the aromatic rings and oxygen. Depending on the FTIR spectral analyses, the bands at 831 cm^{-1}, 916 cm^{-1}, and 1036 cm^{-1} were assigned to the C-O-C, C-O, and C-O-C stretching. The C-H and O-H stretching were observed in the range of 2871–3600 cm^{-1}. In addition, the FTIR spectra of the epoxy resin without hardener appeared to be a valuable tool for both qualitative and quantitative analysis of these processes. It provided the specific information and reaction mechanisms to identify the epoxy resin components and the existence of different vibration bands. Moreover, the O-H band analysis rendered some information about the intermolecular interactions within the components of the epoxy resin.

The sieve analysis result of the river sand is shown in Figure 5.3. The retained percentage of the fine aggregates, the upper and lower limit, was selected according to the Standard Specification for Concrete Aggregates ASTM C33 [30]. In this research, the fine aggregate remains between the upper and lower limit to follow well-graded aggregate. River sand is known to have high silica content—96% of the constituents—acting as filler in the epoxy-modified design mix. The percentage on fine aggregate remains constant in all mix designs.

To evaluate epoxy resin's adequacy as a self-healing agent, five batches of mix designs containing various epoxy resin levels were prepared in the laboratory, Table 5.2. The first batch was prepared without the epoxy resin and considered as the control sample. The control mixture was made by blending the OPC, fine aggregates, and water. In other batches, the epoxy resin to OPC content was varied from 5–20%. To obtain the desired strength properties of the mixes, the quality was controlled strictly during the materials preparation and mixing process by addressing the following criteria:

i. Following to ASTM C1329 and ASTM C109 standards, the ratio of the cement to river sand and water to cement was fixed to 0.33 and 0.48, respectively, for all mix designs.

TABLE 5.2
Mix Proportions of the Self-Healing Mortars

Mix	Cement (kg/m³)	Water (kg/m³)	River Sand (kg/m³)	Epoxy/Cement (%)
1	506	243	1518	0
2	506	243	1518	5
3	506	243	1518	10
4	506	243	1518	15
5	506	243	1518	20

ii. Municipal tap water was added to the concretes during the mixing and curing.
iii. During the preparation stage, a saturated surface dry condition of the river sand was adopted.
iv. During the preparation stage, following [31], first the river sand and OPC were blended together for three minutes, and then the epoxy resin was added to the mix and blended for another two minutes to get a homogenous blend. Subsequently, the water was added to the mix, and blending continued for another three minutes.

The workability of the mortars was measured using the flow table test. After the blend was mixed properly, the fresh mortar was poured onto the molds with two layers, whereby each layer was exposed to 15-second vibration to liberate the air and reduce the pores. The cubical samples with the dimension of 70 mm × 70 mm × 70 mm were used for the compressive strength (CS) test, cylindrical samples of dimension 75 mm × 150 mm were taken for the splitting tensile strength, and the prism specimen of dimension 40 mm × 40 mm × 160 mm was prepared for the flexural strength test. Two types of curing conditions were applied on the obtained specimens, dry and wet-dry. In the wet-dry state, the mortar was immersed in water for five days at (20 ± 3)°C. After five days, the specimens were taken out and left in the ambient conditions (24 ± 3°C) until the testing. In the dry curing, the mortar was left in the ambient state until testing.

5.3 FRESH PROPERTIES

The epoxy-modified mortars' fresh characteristics were evaluated in terms of the workability using the flow table test in compliance with ASTM C230 [32]. The standard conical frustum with a diameter of 100 mm was used for the flow test, where the mortar was placed on the flow table and dropped 25 times within 15 seconds. As the mortar was dropped, it showed a spread out on the flow table, and the diameter of the spread mortar was recorded. This test was repeated three times to calculate the average. The setting times test of the cement pastes was carried out using the Vicat Apparatus according to the ASTM standards [33-34]. In this test, the cement paste was placed into the conical mold of 80 mm diameter and 40 mm height in two layers whereby each

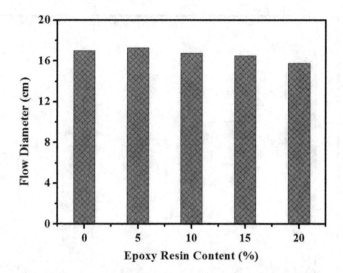

FIGURE 5.4 Effect of epoxy resin content on workability of prepared mortars.

layer was tampered 25 times with a rod. Then, the mortar paste was cured at room temperature, and at every interval of 15 minutes, the specimen was taken out and placed on the Vicat Apparatus to measure the initial and final setting time.

The effects of the epoxy resin on the workability of the prepared mortars were measured using setting time and the flow table test. Figure 5.4 and Figure 5.5 illustrated the values of the measured flow diameter and final sitting time for all tested specimens containing different percentages of epoxy resin. The results indicated that the flow diameters of the fresh mortar were in the range of 169 ± 3 mm, and by increasing the epoxy resin content, the flow diameter tended to decrease as a result of its high viscosity (Figure 5.4). Similarly, the final sitting time was significantly decreased in a modified mortar containing epoxy resin compared to a conventional one (Figure 5.5). For example, the sitting time was decreased by around 70% in the specimen containing 20% epoxy resin compared to the control specimen. Such a phenomenon can be explained by the fact that the mortar mixes became very sticky with the rising levels of the epoxy resin, shortening the setting time of the modified mortar mixture.

5.4 STRENGTH PROPERTIES

The CS of specimens were measured at the age of 28 days using the universal testing machine (UTM) with a 0.3 N/mm²/s loading rate in compliance with ASTM C109 [35]. Similarly, the flexural and splitting tensile strength of epoxy-modified mortars were evaluated in compliance with ASTM C78 [36] and ASTM C496 [37], respectively, after the curing age of 28 days.

For the water absorption test, first three cubical epoxy-modified specimens were submerged inside water for 72 hours (Wts) and after that, oven-dried for 24 hours

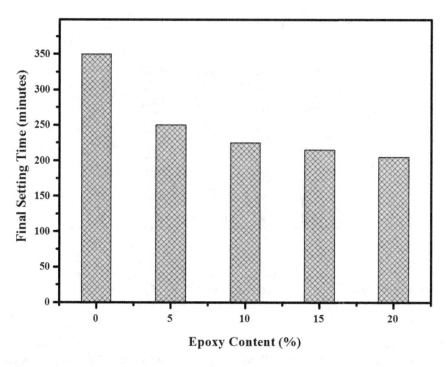

FIGURE 5.5   Final sitting time of all tested specimens.

at 105°C (Wtd). Each sample's weight was measured and percentage of the water absorption (WA) was calculated using the following equation in compliance with ASTM D 6532.

$$WA = \left[(Wts - Wtd)/Wtd \right] \times 100 \qquad (5.1)$$

Wts and *Wtd* (in kg) denote the weight of the treated mortar unit after the immersion in water and oven-dried.

Figure 5.6, Figure 5.7, and Figure 5.8 show the mechanical properties of epoxy resin--modified mortar at 28 days of dry curing age. Three specimens were tested for each mixture and average value considered for compressive strength (CS), flexural strength (FS), and splitting tensile strength (TS). The results indicate that the mechanical properties were improved by an average of 3% and 9% with increasing epoxy resin from 0% to 5% and 0% to 10%, respectively. The highest mechanical properties were achieved by a specimen containing 10% epoxy resin with CS of 33.6 MPa, FS of 2.8 MPa, and splitting TS of 3.3 MPa with standard deviation (SDV) ± 1.6, ± 0.12 and ± 0.22, respectively. Improving the mechanical properties in specimens with 5% and 10% of epoxy resin attributed to the presence of OH- ions from the hydration of Ca(OH)$_2$. Meanwhile, the results confirm that the mechanical properties tend to decrease significantly by increasing epoxy resin content to from 10% to 15% and 20%. For instance, the CS declined by 54% by increasing the epoxy resin from 10% to 20%. The previous

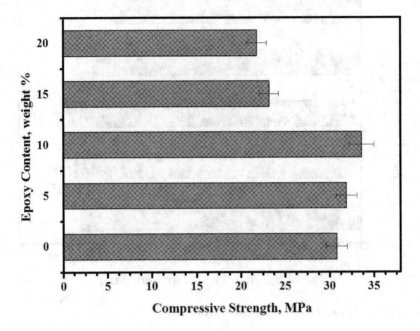

FIGURE 5.6 Compressive strength of all tested specimens.

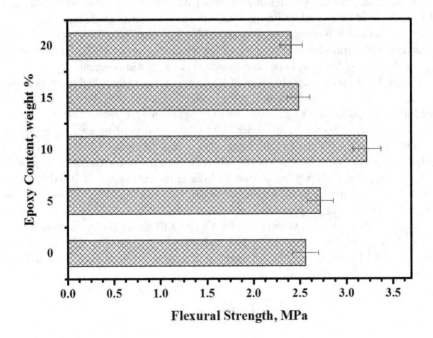

FIGURE 5.7 Flexural strength performance of all tested specimens.

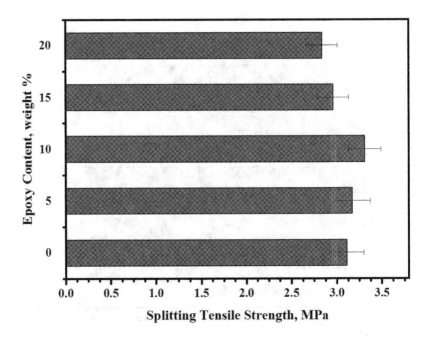

FIGURE 5.8   Tensile strength of all tested specimens.

literature indicated that by increasing the epoxy content by more than 10% in the mix design, the residual unhardened epoxy within the mortar matrix might have interrupted the hydration and polymerization processes [38-40]. It was also acknowledged that the mechanical properties in epoxy resin-modified specimens were increased by an average of 8% under wet-dry curing condition regardless of epoxy resin content.

Figure 5.9 shows the development of CS in epoxy resin–modified mortars at different ages using dry and wet-dry curing conditions. The average values of three readings were adopted for each test with SDV between ±1.4 and ±1.8. It is clear that the CS was increased for all specimens by increasing the curing age. Nevertheless, the increase rate was substantially higher in specimen containing 10% epoxy resin cured under wet-dry curing conditions whereby the CS was increased from 36 MPa at the curing age of 28 days to around 40 MPa at the curing age of 360 days. Such improvement attributed to the OH- content that improve the hydration process and increasing the C-S-H gel product.

Using multi-linear regression analysis, Figure 5.10 shows the correlation among mechanical properties of epoxy resin–modified specimens after 28-days of curing. The ACI [41] proposed the following correlation for mechanical properties of concrete.

$$TS = 0.55\sqrt{CS} \tag{5.2}$$

$$FS = 0.62\sqrt{CS} \tag{5.3}$$

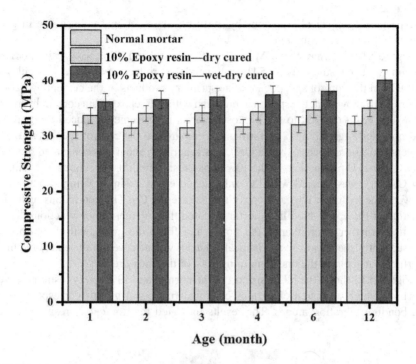

FIGURE 5.9 Effects of curing age on compressive strength development.

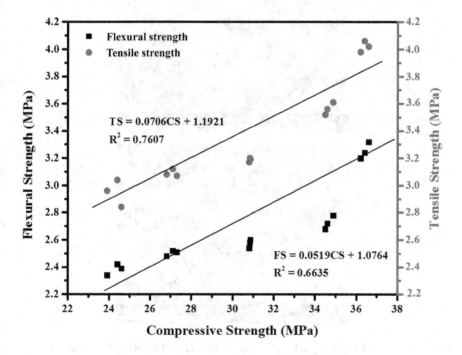

FIGURE 5.10 Correlation between mechanical properties of epoxy resin–modified specimens.

The proposed empirical equations by multi-linear regression analysis are in good agreement with ACI equations.

Figure 5.11 illustrates the SEM image of the control and 10% epoxy resin-modified specimens. The results (Figure 5.11a) indicated the production of C-(A)-S-H gels and Ca(OH)$_2$ in the control specimen that are primary products of the cement hydration. Figure 5.11b shows that the specimen prepared with 10% epoxy resin contain hydroxyl-ion-epoxy resin that led to the microstructure's improvement reduced the porosity and provided high strength performance compared to the control specimen. In the cement hydration process, the Ca(OH)$_2$ was generated, and it was consumed partially over C-(A)-S-H production. In the 10% epoxy resin–modified specimen, the remaining Ca(OH)$_2$ was reacted with the unhardened epoxy resin, creating strong bonds between the hydroxyl ions and epoxy to increase the CS. The continuous hydration reaction in the epoxy-modified mortar initiated the polymerization reactions. These reactions further strengthened the bonds and filled the pores within the mortar. Consequently, the improved mechanical behavior was achieved in the epoxy-modified mortars, particularly the one containing 10% of the epoxy resin.

Figure 5.12 shows the FTIR spectra of the normal and 10% epoxy resin–modified mortars, which consisted of various chemical functional groups corresponding to the bonding vibration modes. The results indicated that the epoxy resin–modified

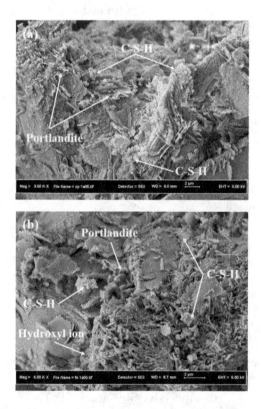

FIGURE 5.11   SEM micrographs of the (a) control and (b) 10% epoxy-based mortar.

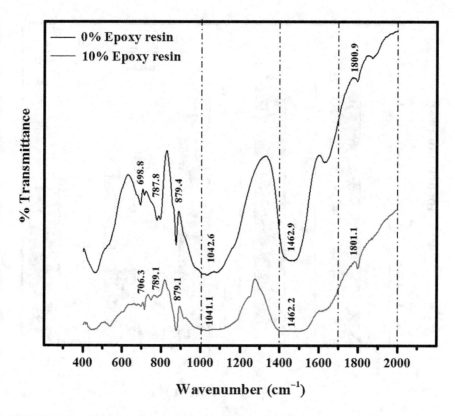

FIGURE 5.12   FTIR spectra analysis of the OPC and 10% epoxy resin-modified mortar.

specimen has a higher $Ca(OH)_2$ absorption due to residual hydroxyl ions used in the epoxy resin reaction. The FTIR spectrum indicated the Si-O, Al-O, and C-H reaction regimes in the mortar metric. The Si-O and Al-O bond vibration energy levels were increased from 698.8 and 787.8 cm^{-1} to 706.3 and 789.1 cm^{-1} with a corresponding rise in the epoxy resin level from 0% to 10%. Increasing demand of the OH- ion and its consumption in the production of the hardened epoxy (hydroxyl-ion-epoxy resin) affected the total amount of $Ca(OH)_2$, which further reduced the consumption of the silica-aluminum in the hydration process. The increase in the bond vibration energy from 875 cm^{-1} to 1045 cm^{-1} was attributed to the generation of the high quantity of $Al(OH)_4$, C-S-H, and C-(A)-S-H gels apart from the hydroxyl-ion-epoxy resin, improving the mechanical performance. The distortion of the bonds in the range of 1100–1800 cm^{-1} might be relevant to the H-O-H's bending vibrations in the H_2O molecules, whereas the C-C ring of the phenol group showed stretching in the epoxy-modified specimen. The addition of 10% of the epoxy resin activated some interactions with the hydroxyl ions, and the reaction of the $Ca(OH)_2$ with the epoxy resin caused the polymerization processes. The amount of $Ca(OH)_2$ was found to decrease as the polymer became reactive.

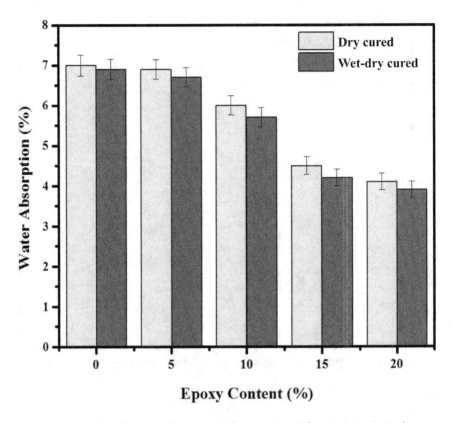

FIGURE 5.13   Effect of epoxy resin content and curing condition on water absorption.

Figure 5.13 shows the effect of various ratios of epoxy resin on water absorption (WA). The average values of three readings were adopted for each mixture. The results indicated that the water absorption is inversely proportional with the epoxy resin content. The WA was declined by around 70% in specimen containing 20% epoxy resin compared to the control specimen. The specimens cured under dry conditions absorb water by average 6% higher compared to wet-dry cure conditions, due to the reaction of the OH- with the unhardened epoxy resin in the modified mortars.

5.5   SELF-HEALING PERFORMANCE

To evaluate the self-healing performance of the studied mortars, the artificial cracks were generated using a compression test machine (loading speed 0.8 kN/s) at the ages of 28, 180, and 360 days. Figure 5.14 illustrates the self-healing evaluation procedure. Such an approach intended to simulate the practical construction scenarios whereby the cracks often appear unexpectedly at different age. At these proposed ages, the average of the developed CS, UPV, porosity, and surface morphology of the specimens were recorded, and then the artificial cracks were generated on specimens by the pre-loading rate of 50% and 80% of their ultimate CS. Subsequently, using

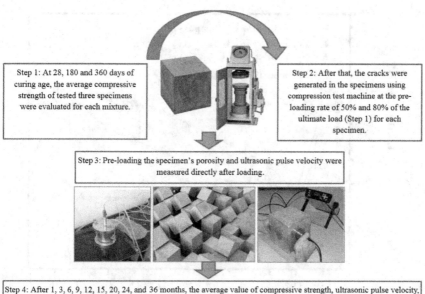

FIGURE 5.14 Self-healing evaluation procedure.

structural morphology tests, specimens were again investigated to ensure that the crack was properly generated inside the mortars, and then the specimens were cured at room temperature 26 ± 3°C until the desired day of testing. The development of the healing was evaluated at each three months' interval (up to 36 months) using mechanical (CS), non-destructive (ultrasonic pulse velocity, or UPV), and structural morphology (SEM) tests.

Ohama [42] recommended that the UPV test can characterize the development of the concrete damage. They proposed the following equation based on reducing the UPV, labeling the degree of damage.

$$Degree\,of\,Damage = 1-[V_a/V_b] \tag{5.4}$$

where V_a and V_b are the velocities after and before the peak loading (m/s) in the mortars.

The self-healing efficiency can also be determined using the following equation.

$$self\text{-}healing\ efficiency = [(CS_2 - CS_1)/CS_1] \times 100 \tag{5.5}$$

where CS_1 and CS_2 are the CS before and after the loading.

The self-healing efficiency of epoxy resin on CS and pulse velocity under 50% and 80% pre-loading is presented in Figure 5.15 and Figure 5.16 for control specimen and specimen containing 10% of epoxy resin, cured under wet-dry conditions. In both

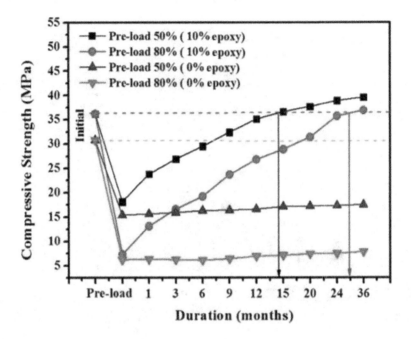

FIGURE 5.15   Self-healing efficiency of epoxy resin on compressive strength.

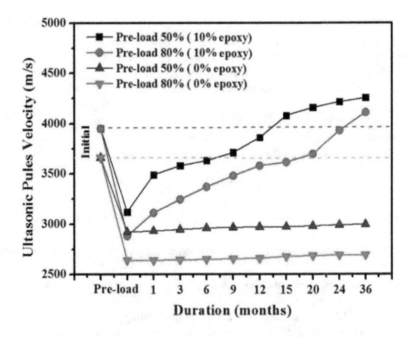

FIGURE 5.16   Self-healing efficiency of epoxy resin on ultrasonic pulse velocity.

specimens, pre-loading was applied at 28 days of curing age. The self-healing function was evaluated by pre-loading mortars by 50% and 80% of maximum load, and the specimens were checked by using the non-destructive test method. Two types of mixtures were adopted to evaluate the self-healing function: the first mixture which was prepared with 0% of epoxy resin and considered as a control sample, and the second mixture was prepared with 10% of epoxy resin and cured in a wet-dry environment. The epoxy-enabled self-healing performance of the modified mortars was examined via the non-destructive and CS tests, whereby self-healing effects were checked in terms of the cracks generation at early curing age. It was found that the cracks were produced at the curing age of 28 days—that is, instantly after the concrete was matured.

For specimens adopted as control sample, the strength was dropped from 30.8 MPa to 15.4 MPa after pre-loading of 50% maximum load. However, the specimens shown slow rate of compressive strength development and the strength recorded 15.6, 15.9, 16.3, 16.4, 16.6, 17.1, 17.2, 17.3, and 17.5 MPa after 1, 3, 6, 9, 12, 15, 20, 24, and 36 months of curing age, respectively. A similar trend was observed with control specimens which were exposed to pre-loading 80% and the very slow rate of strength development was recorded; the average compressive strength after 36 months presented 7.8 MPa compared to 6.2 MPa at the initial pre-loading. Unlike, the specimens containing 10% epoxy resin presented excellent performance compared to control sample. After 50% of pre-loading and reducing the strength from 36.2 MPa to 18.1 MPa, the compressive strength of cured specimens were developed and recorded 23.8, 26.9, 29.5, 32.4, 35.1, 36.6, 37.7, 38.9, and 39.6 MPa after 1, 3, 6, 9, 12, 15, 20, 24, and 36 months, respectively. Specimens pre-loaded with 80% of maximum load showed lower performance compare to that pre-loaded with 50% of maximum load and recorded 13.1, 16.6, 19.2, 23.7, 26.8, 28.9, 31.4, 35.7, and 36.9 MPa for the same periods of curing. Meanwhile, the OPC did not show a noticeable improvement after six months. The ability of the mortars in regaining the initial CS clearly indicated the epoxy stimulated self-healing action. It can be asserted that the use of the epoxy resin in the absence of any hardener as the self-healing mediator is beneficial for the production of the maintenance-free mortar, thereby contributing to sustainable development in the building sectors.

The results indicate that the healing efficiency on CS and UPV is much higher for specimens containing 10% epoxy resin than conventional mortar. The CS in the epoxy-modified specimen significantly improved, from 17.5 MPa after artificial crack generation using 50% pre-load to around 40 MPa over 36 months of healing duration. Almost the same trend was observed for pulse velocity, whereby the velocity in epoxy-modified specimens was increased by an average of 40% after spending 36 months of healing duration. This is an indication that the structure of the specimen is dense, with fewer pores and cracks. The results also acknowledged that the CS and pulse velocity in epoxy resin–modified specimen substantially recovered after generation artificial cracks by 80% pre-load and reached initial condition after spending 36 months of healing duration. On the other hand, a low rate of enhancement was observed by the conventional specimen after 36 months of curing age, which significantly affect its durability performance.

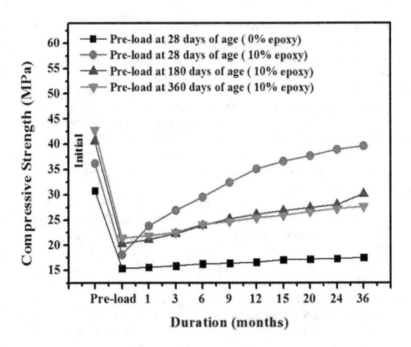

FIGURE 5.17   Effect of artificial cracks age on CS healing efficiency.

Figure 5.17 and Figure 5.18 show the effects of artificial crack generation age on the CS and pulse velocity healing efficiency of epoxy-modified mortar. The results indicate that the CS was recovered to its initial condition once the cracks were generated at the early age of curing, indicating epoxy resin's superior efficiency in healing the cracks. Figure 5.17 shows that the CS was recovered from 17 MPa after generating by 50% pre-load at the curing age of 28 days to more than 40 MPa after 36 months of healing duration, indicating that cracks were completely healed. On the other hand, once the cracks were generated at the later curing age, the healing efficiency declined. For example, once the artificial cracks were generated at 360 days of curing age, the CS only recovered by 35%, from around 20 MPa after generation of artificial cracks to 27 MPa after 36 months' healing duration. The same trend was almost observed for the UPV test, whereby pulse velocity was significantly improved once the artificial cracks were generated at early age of curing. The higher velocity indicates the proper microstructure with fewer pores and cracks. Overall, it was concluded that there is a direct relationship between artificial crack generation and self-healing efficiency, whereby a higher healing efficiency was achieved at a younger crack age. In contrast, the control samples without epoxy after generation of artificial cracks did not recover and did not show any gain in compressive strength.

Figure 5.19 shows the SEM morphology of the self-healing mechanism of epoxy-modified specimens where artificial cracks were generated at 28, 180, and 365 days of curing age. The healing efficiency depends on the age of artificial cracks generation and the amount of the unhardened epoxy resin and $Ca(OH)_2$ available in the

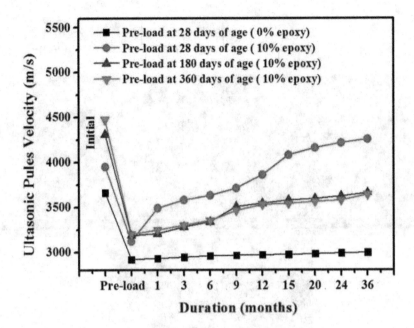

FIGURE 5.18 Effect of artificial cracks age on UPV healing efficiency.

specimen's matrix. After the generation of artificial cracks, the unhardened epoxy started reacting with the OH- from calcium hydroxide that becomes hardened and filling the cracks. Figure 5.19a shows the epoxy-modified specimen's self-healing mechanism whereby artificial cracks were generated at 28 days of curing age. This figure shows that the produced cracks were occupied by the reaction products of the epoxy resin and OH-. Also, the high quantity of the C-S-H gel and Ca(OH)$_2$ were observed from SEM results, whereby they are the main components associated with strength enhancement and healing the cracks.

On the other hand, the SEM morphology shows once the artificial cracks were generated at the later age, (Figure 5.19b and Figure 5.19c), a lower amount of C-S-H gel, Ca(OH)$_2$, and hardened epoxy resin were produced, negatively affected the healing efficiency. In the self-healing mechanism, the OH- generated in the hydration process was significant for producing the C-(A)-S-H gels and self-healing system. The hydroxyl ion was needed for the occurrences of the self-healing action [41-42]. In the bacteria-activated self-healing concretes, the bacterial strain reacts with the oxygen to generate the calcite, further precipitating to heal the cracks. Using a similar mechanism, the epoxy-modified mortars were prepared whereby the epoxy resin reacted with the OH- that enabled the hardening and healing of the cracks. Half of the epoxy resin was hardened, and the remaining one reacted as the healing agent at a later age. The reaction pathway followed the following mechanism:

OH- (From cement hydration) + Unhardened epoxy resin →
Hardened epoxy resin   (5.6)

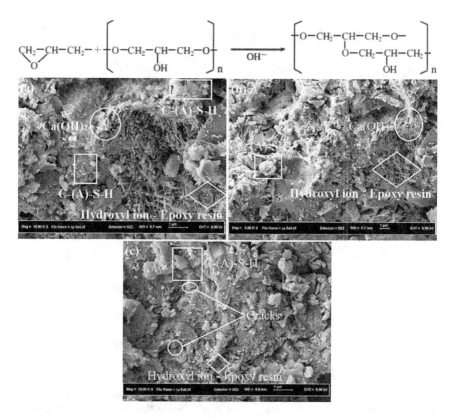

FIGURE 5.19:   SEM morphology of self-healing mechanism of epoxy-modified specimen after artificial cracks generation at (a) 28 days, (b) 180 days, and (c) 365 days of curing age.

Figure 5.20 and Figure 5.21 show the degree of damage and healing efficiency as a measure to evaluate the adequacy of epoxy resin for healing the artificial cracks. Similar to CS and UPV, the degree of damage and healing efficiency is a function of the generation age of artificial cracks. A superior performance was achieved by the epoxy-modified specimen once the artificial cracks were generated at an early age where the degree of damage and healing efficiency surpasses 0% and 100%, respectively. On the other hand, once the artificial cracks were generated at the later curing ages of 180 and 360 days, the healing efficiency was declined by 30% and 50%, respectively. The degree of damage also was recorded 0.15 and 0.2 (from the initial damage of 0.25 and 0.3) in these specimens after spending 36 months of healing duration. Such a figure is equivalent to a 50% and 30% healing efficiency, as shown in Figure 5.21.

5.6   DEVELOPING AN ANN TO ESTIMATE DEGREE OF DAMAGE AND HEALING EFFICIENCY

This section develops an ANN combined with a metaheuristic algorithm to estimate the degree of damage and healing efficiency of studied epoxy-modified specimens. The

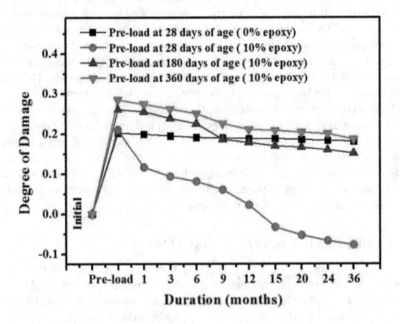

FIGURE 5.20   Degree of damage of epoxy-modified mortars.

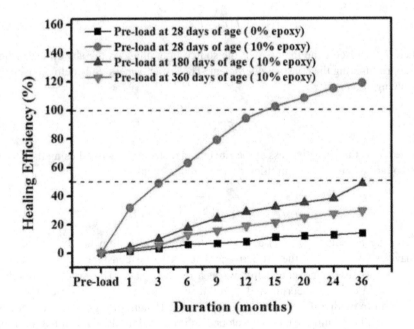

FIGURE 5.21   Degree of damage and Healing efficiency of epoxy-modified mortars.

multilayer feed-forward network provides a reliable feature for the ANN structure, and therefore was used in this research. The multilayer feed-forward network comprises three individual layers: the input layer, whereby the data are defined to the model; the hidden layer/s, whereby the input data are processed; and finally, the output layer, whereby the results of the feed-forward ANN are produced. Each layer contains a group of nodes referred to as neurons that are connected to the proceeding layer. The neurons in hidden and output layers consist of three components; weights, biases, and an activation function that can be continuous, linear, or nonlinear. Standard activation functions include nonlinear sigmoid functions (logsig, tansig) and linear functions (poslin, purelin) [43]. Once the architecture of a feed-forward ANN (number of layers, number of neurons in each layer, activation function for each layer) is selected, the weight and bias levels should be adjusted using training algorithms. One of the most reliable ANN training algorithms is the backpropagation (BP) algorithm, which distributes the network error to arrive at the best fit or minimum error [44–45].

5.6.1 Firefly Optimization Algorithm (FOA)

Fireflies, also known as lightning bugs, are nocturnal, luminous beetles. Several researchers have studied the behavior of this creature in nature [46]. The nature-inspired firefly algorithm is a metaheuristic algorithm proposed by Fister et al. [47] and stimulated by the insect's flashing behavior. In the classic firefly optimization algorithm, two fundamental aspects need to be clarified: the source light and attractiveness. The intensity of light I is referred to as an absolute measure of emitted light by the firefly, while the attractiveness β is the measure of light seen by the other fireflies. The intensity of light is defined using following equation.

$$I(r) = I_0 e^{-\gamma r^2} \tag{5.7}$$

where I_0 represents the intensity of the source light and γ is the absorption of light by approximating the constant coefficient. The attractiveness β is defined using the following equation.

$$\beta = \beta_0 e^{-\gamma r^2} \tag{5.8}$$

where β_0 is the attractiveness at the Euclidean distance γ defined using the following equation between two fireflies s_i and s_j, and is equal to 0:

$$\gamma_{ij} = \left\| s_i - s_j \right\| \sqrt{\sum_{k=1}^{k=n} \left(s_{ik} - s_{jk} \right)^2} \tag{5.9}$$

According to Equations (5.7) and (5.8), the firefly algorithm has two asymptotic behaviors. As $\gamma \to 0$, the attractiveness tends to be constant ($\beta = \beta_0$), and as $\gamma \to \infty$, the firefly movement follows at a random walk. The firefly algorithm is attractive in civil engineering and material sciences. Bui et al. [48] used the firefly algorithm to predict the compressive and tensile strength of high-performance concrete. Sheikholeslami et al. [49] developed an improved firefly algorithm to optimize reinforced concrete retaining walls. Nigdeli et al. [50] developed a firefly algorithm to optimize reinforced concrete footings.

5.6.2 Generation of Training and Testing Data Sets

To train and develop a reliable ANN, the mechanical and material properties of epoxy-modified design mixes were taken into account based on input variables (see Table 5.3).

Since the behavior and number of input data should be statistically evaluated against the output data, Figure 5.22 and Figure 5.23 show the probability plot of output data.

The ANN used in this study is a new feed-forward model. Eighty percent of input data, out of 36 samples, were used for training, and the remaining 20% were considered for testing the network. According to the characteristics of the available input data and the number of outputs, a two-layer ANN was proposed in the initial attempt, and its adequacy was evaluated by using several measures. Therefore, the trial and error method is used to obtain the ideal architecture, the architecture that best reflects the characteristics of the laboratory data. In this research, an innovative method for calculating the number of neurons in hidden layers was taken into account, as shown in the following equation.

$$N_H \leq 2N_I + 1 \tag{5.10}$$

where N_H is the number of neurons in the hidden layers and N_I is the number of input variables.

Since the number of effective input variables is six, the empirical equation shows that the number of neurons in hidden layers can be less than 13. Therefore, several networks with different topologies—with a maximum of two hidden layers and a maximum of 13 neurons—were trained and studied in this study. The hyperbolic tangent stimulation function and Levenberg–Marquardt training algorithm were used

TABLE 5.3
Characteristics of Studied Input and Output Parameters

Num.	Parameter	Type	Unit	Max.	Min.	Average	STD
1	Epoxy	Input	kg/m³	50.6	0	37.95	22.22
2	Pre-load at age	Input	(Day)	360	28	149	138.65
3	Healing duration	Input	(Month)	36	1	14	10.71
4	CS	Input	(MPa)	39.6	15.6	25.2	6.79
5	UPV	Input	(Km/s)	4.26	2.93	3.45	0.37
6	Water absorption	Input	(%)	14.8	5.5	9.57	3.08
7	Observation degree of damage	Output		0.27	-0.08	0.16	0.09
8	Observation healing efficiency	Output		118.78	0	34.04	34.7

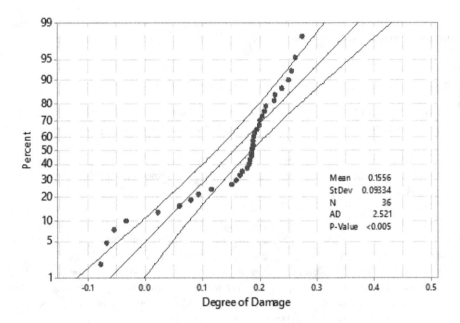

FIGURE 5.22: Probability plot for output variables (a) degree of damage.

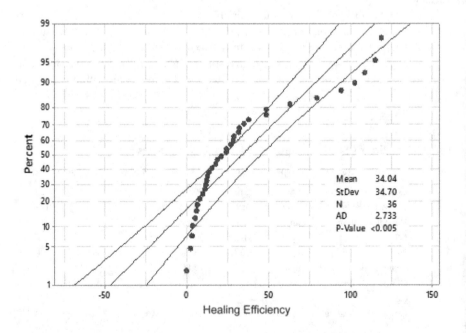

FIGURE 5.23 Probability plot for output variables healing efficiency.

in all networks. The statistical indices used to evaluate the performance of different topologies are root mean squared error (RMSE), average absolute error (AAE), model efficiency (EF), and variance account factor (VAF), which are defined as follows.

$$MSE = \frac{1}{n}\sum_{i=1}^{n}\left(P_i - O_i\right)^2 \tag{5.11}$$

$$ME = \frac{1}{n}\sum_{i=1}^{n}\left(P_i - O_i\right) \tag{5.12}$$

$$MAE = \frac{1}{n}\sum_{i=1}^{n}\left|P_i - O_i\right| \tag{5.13}$$

$$RMSE = \left[\frac{1}{n}\sum_{i=1}^{n}\left(P_i - O_i\right)^2\right]^{\frac{1}{2}} \tag{5.14}$$

After examining different topologies, it was found that the network with 6-6-4-2 topology has the lowest value of error in RMSE, AAE, EF, and VAF, and the highest value of R^2, to estimate the two output parameters as shown in Table 5.4. It is necessary to mention that the error criteria for training and testing the data are calculated in the main range of variables and not in the normal range.

Figure 5.24 shows the topology of a feed-forward network with two hidden layers, six input variables (neurons), and two output parameters.

To optimize the ANN's weights and biases, the FOA has been used to provide the least prediction error for trained structure. The properties of the FOA parameters are shown in Table 5.5.

5.6.3   RESULTS

The results of the FOA-ANN models are shown in Figure 5.25, Figure 5.26, Figure 5.27, and Figure 5.28 for the degree of damage and healing efficiency output parameters, respectively. The results indicate that the FOA-ANN estimated a reliable result for the ratio of observational to computational values, R^2, for both input parameters, indicating the proposed model's high potential and accuracy.

Table 5.6 provides the final weights and biases for both hidden layers estimated by the FOA-ANN model. Using the values of the weights and biases between the

TABLE 5.4
Statistical Indices of ANN with 6-6-4-2 Topology Combined with FOA

Training				Testing			
MSE	ME	MAE	RMSE	MSE	ME	MAE	RMSE
2.94	0.26	0.92	1.71	3.72	0.89	1.11	1.93

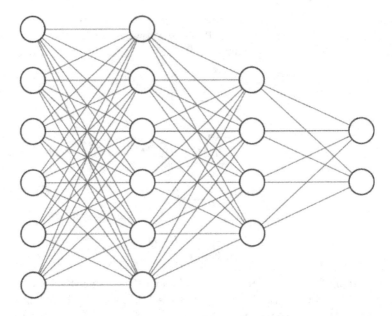

FIGURE 5.24 The topology of a feed-forward with two hidden layers (6-6-4-2).

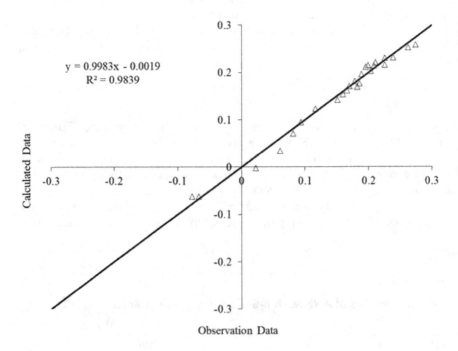

FIGURE 5.25 Predicted versus experimental values of degree of damage output using training data.

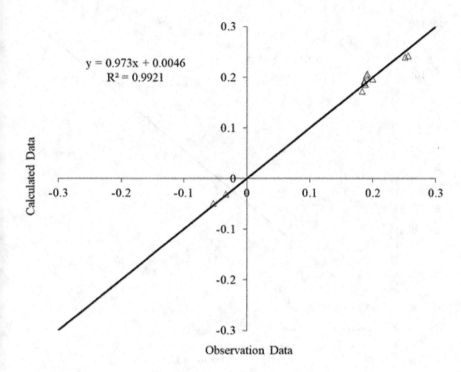

FIGURE 5.26 Predicted versus experimental values of degree of damage output using testing data.

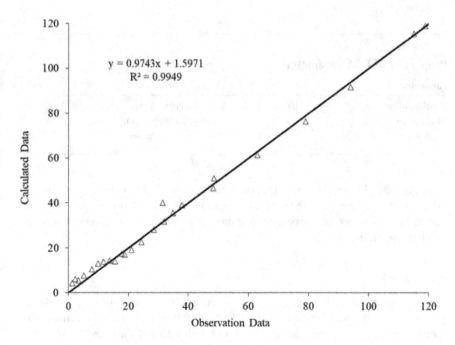

FIGURE 5.27 Predicted versus experimental values of healing efficiency output using training data.

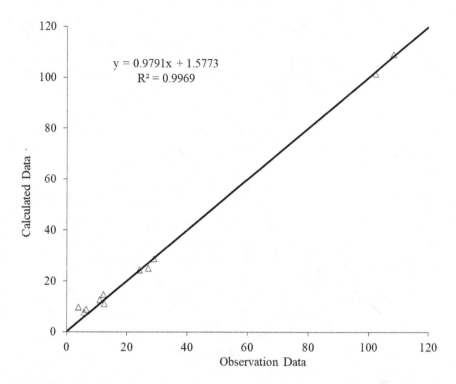

FIGURE 5.28 Predicted versus experimental values of healing efficiency output using testing data.

TABLE 5.5
Properties of FOA Parameters

Parameter	Value	Parameter	Value
Population size	100	Attraction coefficient base value	2
Mutation coefficient	0.25	Mutation coefficient damping ratio	0.99
Light absorption coefficient	1	M (exponent of distance term)	2

different ANN layers, the two output parameters (degree of damage and healing efficiency) can be determined and predicted. Besides, these final weights and bias values can be used to design an epoxy-modified mortar with targeted mechanical properties and healing efficiency.

5.7 SUMMARY

This chapter investigated the application of epoxy resin polymer as a self-healing strategy for improving the mechanical and durability properties of cement-based mortar. Several mixes were designed, with various ratios of the epoxy resin to OPC,

TABLE 5.6

Final Weights and Bias Values of the Optimum FA-ANN Model 6-6-4-2

IW						b1
−0.2226	0.6913	0.7234	0.4921	−0.2198	0.7664	0.3660
0.5876	0.2296	0.0543	−0.9068	−0.8038	0.7919	−0.4297
0.8156	0.8708	0.2050	−0.7444	−0.4160	−0.3075	0.9485
0.3177	−0.3422	−0.9205	−0.8923	0.5018	−0.0903	−0.4386
0.0142	−0.9034	−0.3105	0.9778	0.5866	0.3670	−0.0370
−0.7485	−0.0284	−0.7803	−0.2927	0.7430	−0.8874	−0.8910

LW1						b2
0.1391	−0.6483	−0.9244	−0.2580	0.3317	−0.2730	−0.5625
0.1451	−0.9574	0.9974	0.7265	−0.0867	−0.6183	0.5001
−0.0218	0.1769	−0.2386	0.5438	−0.2493	−0.0382	−0.8893
0.3847	−0.1255	−0.1702	0.4274	−0.7884	−0.6989	0.2426

LW2						b3
−0.4282	−0.0805	0.6376	0.7041			0.0395
0.7643	0.4558	−0.4930	−0.4541			−0.3842

IW: Weights values for input layer
LW1: Weights values for first hidden layer
LW2: Weights values for second hidden layer
b1: Bias values for first hidden layer
b2: Bias values for second hidden layer
b3: Bias values for output layer

to assess the efficiency of the epoxy as the self-healing agent. The mechanical properties, UPV, and SEM morphologies were recorded to determine the feasibility of implementing the epoxy resin as the self-healing agent in mortar manufacturing. In addition, by using the available experimental test database, an optimized ANN model with combined with the firefly optimization algorithm (FOA-ANN) was developed to estimate the degree of damage and healing efficiency of mix designs. The following provides the main findings of this chapter:

i. The final setting time and water absorption were significantly decreased by increasing the epoxy resin content in the modified mortar compared to a conventional specimen.

ii. The highest mechanical properties were achieved by a specimen containing 10% epoxy resin with that attributed to the presence of OH- ions from the hydration of $Ca(OH)_2$. However, the mechanical properties significantly decreased by increasing epoxy resin content from 10% to 15% and 20%. Such phenomenon can explain by the residual unhardened epoxy within the mortar matrix that may interrupted the hydration and polymerization processes.

iii. The SEM images indicated that the specimen prepared with 10% epoxy resin contained hydroxyl-ion-epoxy resin that led to the microstructure's improvement, reduced the porosity, and providing high strength performance compared to the control specimen. In addition, in this specimen the $Ca(OH)_2$ was reacted with the unhardened epoxy resin, creating strong bonds between the hydroxyl ions and epoxy to increase the CS.

iv. The healing efficiency on CS and UPV is much higher for the specimen containing 10% epoxy resin than conventional mortar. Besides, it was concluded that there is a direct relationship between artificial crack age and self-healing efficiency, whereby a higher healing efficiency was achieved at a younger crack age. Similarly, the degree of damage and healing efficiency were recorded 0% and 100%, respectively, for epoxy-modified specimens once the artificial cracks were generated at an early age.

v. The ANN combined with the metaheuristic firefly algorithm provided satisfactory results to estimate the degree of damage and healing efficiency in epoxy-modified specimens. Additionally, the firefly algorithm optimization can be used as a powerful tool in optimizing ANN weights. By using the optimized weight and bias of FOA-ANN, it is possible to design mixes with targeted degree of damage and healing efficiency, depending on particular environments.

REFERENCES

1. Shah, K.W., and G.F. Huseien, *Biomimetic self-healing cementitious construction materials for smart buildings.* Biomimetics, 2020.**5**(4): p. 47.
2. Batis, G., P. Pantazopoulou, S. Tsivilis, and E. Badogiannis, *The effect of metakaolin on the corrosion behavior of cement mortars.* Cement and Concrete Composites, 2005.**27**(1): p. 125–30.
3. Triantafillou, T.C., and C.G. Papanicolaou, *Shear strengthening of reinforced concrete members with textile reinforced mortar (TRM) jackets.* Materials and Structures, 2006.**39**(1): p. 93–103.
4. Xu, J., and W. Yao, *Multiscale mechanical quantification of self-healing concrete incorporating non-ureolytic bacteria-based healing agent.* Cement and Concrete Research, 2014.**64**: p. 1–10.
5. Wu, M., B. Johannesson, and M. Geiker, *A review: Self-healing in cementitious materials and engineered cementitious composite as a self-healing material.* Construction and Building Materials, 2012.**28**(1): p. 571–583.
6. Edvardsen, C., Water permeability and autogenous healing of cracks in concrete. In *Innovation in concrete structures: Design and construction.* 1999: Thomas Telford Publishing, Scotland, UK. p. 473–487.
7. Ait Ouarabi, M., et al., *Ultrasonic monitoring of the interaction between cement matrix and alkaline silicate solution in self-healing systems.* Materials, 2017.**10**(1): p. 46.
8. Xue, C., et al., *Effect of incompatibility between healing agent and cement matrix on self-healing performance of intelligent cementitious composite.* Smart Materials and Structures, 2020.**29**(11): p. 115020.
9. Lee, Y.-S., and J.-S. Ryou, *Crack healing performance of PVA-coated granules made of cement, CSA, and Na2CO3 in the cement matrix.* Materials, 2016.**9**(7): p. 555.

10. Qiu, J., S. He, and E.-H. Yang, *Autogenous healing and its enhancement of interface between micro polymeric fiber and hydraulic cement matrix.* Cement and Concrete Research, 2019.**124**: p. 105830.

11. Yıldırım, G., et al., *A review of intrinsic self-healing capability of engineered cementitious composites: Recovery of transport and mechanical properties.* Construction and Building Materials, 2015.**101**: p. 10–21.

12. Jonkers, H.M., et al., *Application of bacteria as self-healing agent for the development of sustainable concrete.* Ecological Engineering, 2010.**36**(2): p. 230–235.

13. Lv, L.-Y., et al., *Experimental and numerical study of crack behaviour for capsule-based self-healing cementitious materials.* Construction and Building Materials, 2017.**156**: p. 219–229.

14. Minnebo, P., et al., *A novel design of autonomously healed concrete: Towards a vascular healing network.* Materials, 2017.**10**(1): p. 49.

15. Seifan, M., et al., *The effect of cell immobilization by calcium alginate on bacterially induced calcium carbonate precipitation.* Fermentation, 2017.**3**(4): p. 57.

16. Gollapudi, U., et al., *A new method for controlling leaching through permeable channels.* Chemosphere, 1995.**30**(4): p. 695–705.

17. Ohama, Y., K. Demura, and T. Endo, *Strength properties of epoxy-modified mortars without hardener.* Proceedings of the 9th international congress on the chemistry of cement, 1992.

18. Ohama, Y., and V. Ramachandran, Polymer-modified mortars and concretes. In *Concrete admixtures handbook (second edition).* 1996: Elsevier, Netherlands. p. 558–656.

19. May, C., *Epoxy resins: Chemistry and technology.* 1987: CRC press, New York, NY. p. 1–102.

20. Ariffin, N.F., M.W. Hussin, and M.A. Rafique, *Properties of polymer-modified mortars.* 2015: UTM Press, Johor Bahru, Malaysia. p. 1–9.

21. Ariffin, N.F., et al., *Strength properties and molecular composition of epoxy-modified mortars.* Construction and Building Materials, 2015.**94**: p. 315–322.

22. Ohama, Y., K. Demura, and T. Endo, Properties of polymer-modified mortars using epoxy resin without hardener. In *Polymer-modified hydraulic-cement mixtures.* 1993: ASTM International, ASTM, Michigan, United States. p. 1–7.

23. Ariffin, N.F., et al., Degree of hardening of epoxy-modified mortars without hardener in tropical climate curing regime. In *Advanced materials research.* 2015: Trans Tech Publ, Switzerland, **1113**. p. 28–35.

24. Bhutta, M.A.R., *Effects of polymer—cement ratio and accelerated curing on flexural behavior of hardener-free epoxy-modified mortar panels.* Materials and Structures, 2010.**43**(3): p. 429–439.

25. Jo, Y., *Basic properties of epoxy cement mortars without hardener after outdoor exposure.* Construction and Building Materials, 2008.**22**(5): p. 911–920.

26. Łukowski, P., and G. Adamczewski, *Self-repairing of polymer-cement concrete.* Bulletin of the Polish Academy of Sciences: Technical Sciences, 2013.**61**(1): p. 195–200.

27. ASTM, *ASTM C150: Standard specification for Portland cement.* 2001: ASTM, Philadelphia, PA.

28. Sitarz, M., W. Mozgawa, and M. Handke, *Vibrational spectra of complex ring silicate anions—method of recognition.* Journal of Molecular Structure, 1997.**404**(1–2): p. 193–197.

29. González, M.G., J.C. Cabanelas, and J. Baselga, *Applications of FTIR on epoxy resins-identification, monitoring the curing process, phase separation and water uptake.* Infrared Spectroscopy-Materials Science, Engineering and Technology, 2012.**2**: p. 261–284.

30. ASTM, C., *Standard specification for concrete aggregates.* 2003: American Society for Testing and Materials, Philadelphia, PA.

31. Issa, M.A., Effect of Portland Cement (current ASTM C150/AASHTO M85) with limestone and process addition (ASTM C465/AASHTO M327) on the performance of concrete for pavement and bridge decks, 2014, Illinois center for transportation.

32. ASTM, A., *Standard specification for flow table for use in tests of hydraulic cement.* 2014: ASTM, West Conshohocken, PA.

33. ASTM, C., *Standard test methods for time of setting of hydraulic cement by Vicat needle.* 2008. ASTM, Michigan, United States. p. 1–8.

34. ASTM, C., Standard test method for flexural strength of concrete (using simple beam with third-point loading). In *American society for testing and materials.* 2010. ASTM, Michigan, United States. p. 1–6.

35. ASTM, A., *ASTM C496/C496M-04e1 standard test method for splitting tensile strength of cylindrical concrete specimens.* Annual Book of ASTM Standards 2008: Section, 2008.**4**.

36. ASTM-C496, *Standard test method for splitting tensile strength of cylindrical concrete.* 1996: American Society for Testing and Materials, West Conshohocken, PA.

37. (ASTM), A.S.f.T.a.M., *Standard test method for evaluation of the effect of clear water repellent treatments on water absorption of hydraulic cement mortar specimens.* ASTM D6532, 2014. ASTM, Michigan, United States. p. 1–10.

38. Suaris, W., and V. Fernando, *Ultrasonic pulse attenuation as a measure of damage growth during cyclic loading of concrete.* ACI Materials Journal, 1987.**84**(3): p. 185–193.

39. Committee, A., and I.O.F. Standardization, *Building code requirements for structural concrete (ACI 318–08) and commentary.* 2008: American Concrete Institute, Michigan, United States. p. 1–459.

40. Standard, A., *ASTM C109-standard test method for compressive strength of hydraulic cement mortars.* 2008: ASTM International, West Conshohocken, PA.

41. Huseien, G., et al., *Effect of binder to fine aggregate content on performance of sustainable alkali activated mortars incorporating solid waste materials.* Chemical Engineering Transactions, 2018.**63**: p. 667–672.

42. Ohama, Y., *Handbook of polymer-modified concrete and mortars: Properties and process technology.* 1995: William Andrew, Elsevier, Netherlands. p. 1–236.

43. Nikoo, M., et al., *Determining the natural frequency of cantilever beams using ANN and heuristic search.* Applied Artificial Intelligence. Taylor and Francis, Florida, 2018.**32**(3): p. 309–334.

44. Haykin, S., *Neural networks: A comprehensive foundation.* 2007: Prentice-Hall, Inc. Cambridge University, Cambridge, UK. p. 409–412.

45. Bishop, C.M., *Pattern recognition and machine learning.* 2006: Springer, Switzerland. p. 1–46.

46. Gandomi, A.H., X.-S. Yang, and A.H. Alavi, *Mixed variable structural optimization using firefly algorithm.* Computers & Structures, 2011.**89**(23–24): p. 2325–2336.

47. Fister, I., et al., *A comprehensive review of firefly algorithms.* Swarm and Evolutionary Computation, 2013.**13**: p. 34–46.

48. Bui, D.-K., et al., *A modified firefly algorithm-artificial neural network expert system for predicting compressive and tensile strength of high-performance concrete.* Construction and Building Materials, 2018.**180**: p. 320–333.

49. Sheikholeslami, R., et al., *Optimization of reinforced concrete retaining walls via hybrid firefly algorithm with upper bound strategy.* KSCE Journal of Civil Engineering, 2016.**20**(6): p. 2428–2438.

50. Nigdeli, S.M., G. Bekdaş, and X.-S. Yang, *Metaheuristic optimization of reinforced concrete footings.* KSCE Journal of Civil Engineering, 2018.**22**(11): p. 4555–4563.

6 Bacteria-Based Self-Healing Concrete

6.1 INTRODUCTION

Concrete is one of the most prevalently used materials in buildings and construction. Concrete crack formations are unavoidable, as while concrete possesses high compressive strength, it is weak in tension. The lifespan of concrete structures are often decreased due to the presence and development of crack formations. Several procedures that exact high costs and substantial amounts of time can be enacted to repair cracked concrete [1]. Contrastingly, creating self-healing concrete is a more measured approach that leads to the concrete autonomously healing. Upon mixing the concrete, bacteria with calcium nutrient sources are incorporated. As a result, calcium carbonate precipitate occurs upon the formation of a crack, which in turn fills in the crack formation. Standard concrete has less strength than bacterial concrete, demonstrating how calcite precipitation—the basis of the biotechnological approach—enhances the durability and strength of concrete structures [2–3].

Bacteria that trigger calcite precipitation to autonomously heal crack formations find it challenging to plug in crack formations larger than 0.8 mm [4–5]. Bacteria-based mortar suffers a decrease in strength when fine aggregate leads are replaced by lightweight aggregates. Standard lightweight mortar has less strength than bacterial lightweight mortar, demonstrating the latter's utility in lightweight structures. Durability and healing efficacy of a structure are enhanced by lightweight aggregates that are ideal vehicles for bacteria [6–7]. Regardless of age, calcite precipitation leads to significant improvements to concrete strength such as in the case when bacteria was combined with rice husk ash concrete [8–9]. When calcium carbonate precipitation levels are at their highest, M50 grade concrete has its strength properties improved by up to 24% [10].

The permeability and porosity of fly ash concrete are decreased upon the incorporation of *Sporoscarcina pasteurii* bacteria while improvements to the strength occur. Standard concrete experiences four times less water absorption than the mentioned mix, and the compressive strength of the fly ash concrete can rise by up to 22% [11]. In the initial stages of a crack forming, the mentioned autonomous healing technique works to fill in the crack, demonstrating its potential [12]. Nevertheless, the literature demonstrates how the autonomous healing approaches of vascular systems, capsules, and hydro gel encapsulation that trigger calcite precipitation in concrete specimens are sufficiently capable of self-healing concrete cracks. Incorporating cementitious materials in concrete as a potential autonomous healing mechanism is illustrated in Figure 6.1. Calcite precipitation triggered by the bacteria can come for several distinct calcium sources. The latest developments in nano- and biotechnology are utilized to enhance the durability and general properties of concrete.

DOI: 10.1201/9781003195764-6

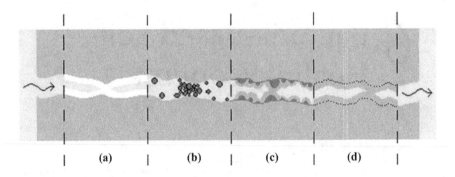

FIGURE 6.1   Possible self-healing mechanisms for cementitious materials.

6.2   SELF-HEALING APPROACH

The secretion of the healing agent would be triggered after detecting a crack formation or damage within an optimal autonomously healing system. The restoration of micro-crack formations in concrete are adeptly handled by autonomous healing techniques. Cracks located on the concrete's surface are efficiently dealt with by autonomous healing approaches. Calcium carbonate precipitation is molded into a pervious layer on the cracks within the concrete, following the incorporation of the bacteria [13–14]. Resistance to alkali environments is a characteristic of the incorporated bacteria, as concrete is a very alkaline material [15–16]. The binding of gravel and sand in concrete is facilitated by the microbiological induction of calcium carbonate precipitation that also aids the plugging in of micro-crack formations [17]. The durability of concrete can be enhanced by the incorporation of microorganisms in the calcite precipitation. Within a highly alkaline environment, *Bacillus sphaericus* can precipitate calcium carbonate via the conversion of urea into carbonate and ammonium [18]. The concrete is capable of self-healing crack formations smaller than 0.2 mm in size. The entrance of deleterious materials is not halted when the crack formations surpass 0.2 mm in size, as the self-healing properties are not sufficient. The hibernation stage of the bacteria within the concrete is disrupted upon the appearance of crack formations in autonomously healing concrete. The crack formations are filled when autonomously healing as the calcium carbonate precipitates into them due to the metabolic activities of bacteria. The hibernation stage recommences once calcium carbonate plugs in the crack formations entirely. Resultantly, any new crack formations lead to the activation of the bacteria and the plugging of such formations. Microbiologically induced calcium carbonate precipitation (MCIP) is a mechanism describing the nature of bacteria as a long-term healing agent.

As seen in Table 6.1, the metabolic pathways of multiple bacteria determine the formation of calcium carbonate. Autotrophic processes are said to generate less precipitated calcium carbonate than heterotrophic processes. The requirement for organic carbon sources for growth is characteristic of heterotrophs that are organisms incapable of using carbon to make their own organic compounds. Contrastingly, simple substances are used to generate complex organic compounds, in the case of

TABLE 6.1
Different Metabolic Pathways of Bacterial Calcium Carbonate Precipitation [19]

Autotrophic Bacteria		Heterotrophic Bacteria			
Non-methylotrophic methanogenesis	Assimilatory pathways	Dissimilatory pathways			
	Urea decomposition	Oxidation of organic carbon			
Anoxygenic photosynthesis		Aerobic		Anaerobic	
		Process/e-acceptor		Process/e-acceptor	
Oxygenic photosynthesis	Ammonification of amino acids	Respiration	O_2	NOx reduction	NO_3/NO_2
		Methane oxidation	CH_4/O_2	Sulfate reduction	SO_4^2

autotroph organisms such as chemosynthesis and photosynthesis. The bacterial cells are protected by a mineral layer over them, following calcium carbonate being generated by the microbials [19–20].

Urea lysis leads to the generation of calcium carbonate precipitate, and multiple microorganisms are capable of the feat. Specific implementations of bacteria were demonstrated via a comprehensive literature review. Graphite nanoplatelets in conjunction with lightweight aggregate in concrete can lead to enhanced strength following the addition of *Bacillus subtilis* [3]. Improvements to the durability of concrete were noted when rice husk ash concrete had the *Bacillus aerius* bacteria added to it [9]. An increase almost by a quarter in compressive strength was observed when the concrete had *Bacillus megaterium* bacteria incorporated into it [10]. Additionally, the durability of concrete increased via the addition of *Bacillus sphaericus* into concrete and the resulting calcium carbonate precipitate [21]. Following autonomous healing, the durability and strength of fly ash concrete was found to increase due to the added *Sporoscarcina pasteurii* bacteria [11]. Similarly, the same bacteria had an equivalent effect in silica fume concrete [11]. Another option for surface treatment of concrete was discovered via the introduction of *Bacillus sphaericus* into concrete with the purpose of monitoring surface treatment [22].

Encapsulation and direct application make up the two principal approaches for applying the healing agent to the concrete, according to the literature. Out of the methods for applying the healing agent to the concrete, which include graphite nanoplatelets (GNP), addition of bacteria in light weight aggregates (LWA), and the direct method, the first method was found to produce the best outcomes when autonomously healing crack formations, demonstrating GNP's proficiency as a carrier compound for bacteria [3]. The best concentration of bacteria was found to be 30 × 105 cfu/ml after determining said quantity and the resultant strength via the direct method of applying the healing agent [10]. The general autonomous healing

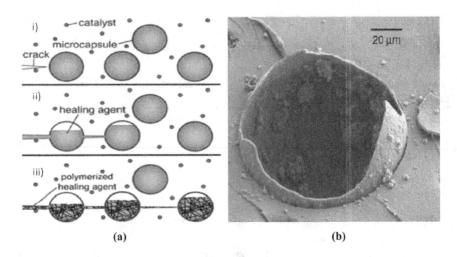

FIGURE 6.2 (a) Simple process of microcapsule approach: (i) formation of cracks in matrix; (ii) process of releasing healing agent; (iii) process of crack healing; and (b) ESEM image displaying a ruptured microcapsule [1].

performance of concrete is enhanced via a different method that involves encapsulation in a polymer-based coating layer after lightweight aggregates are impregnated by a bacteria solution [22]. The autonomous healing of materials via the microencapsulation approach is illustrated in Figure 6.2a. Capillary movement is utilized by the healing agent as it is secreted into crack formation following the embedded microcapsules being broken. Crack formations in close proximity also undergo self-repair and polymerization as the embedded catalyst act with the healing agent. A standard broken microcapsule is illustrated in Figure 6.2b [23]. Quicker detection and reactions to a crack in the concrete matrix alongside the ability to self-heal more kinds of cracks demonstrate the proficient autonomous healing that the encapsulation method offers [24]. In terms of the autonomous healing of cracks and the quantity of precipitation in hydro gel–encapsulated bacterial spores, the efficacy of self-healing was enhanced via the hydro gel encapsulation method [21]. The compressive strength of cement mortar increased by a quarter in almost a month when the *Shewanella* bacteria was added to the concrete via the direct method application approach [25]. The safeguarding of bacteria in the highly alkaline environment alongside its uniform distribution enable the encapsulation method to more effectively produce calcium carbonate precipitation and fill in more crack formations, as per the findings in the literature.

6.3 EFFECT OF BACTERIA

The calcium source supplied determines whether the setting time of the concrete quickens or slows down upon the incorporation of bacteria spore powder. Calcium formatted, calcium nitrate, and calcium lactate are the nutrients provided for the bacteria. The setting time of concrete quickens upon adding calcium nitrate and calcium formatted, while it slows down upon incorporating calcium lactate [5, 26].

Calcite precipitation serves as the basis for a biotechnological approach that enhances the strength of concrete structures. Due to the permeability of the cement mortar, the opening curing period was highly nourishing for the microbial cells acclimatizing to a novel environment. During the curing period, the high pH conditions of the cement can be adapted, which means that the bacterial cells grow gradually. Multiple ions located in the media can lead to calcite precipitates forming in the cement mortar matrix and the surface of the cell, while the cells continue to grow. Consequently, the cement mortar experiences a reduction in permeability and porosity. The simultaneous filling in of multiple pores in the matrix leads to the halting of the flow of oxygen and nutrients to the bacterial cells. As a result, endospores or death are the outcome for these cells. This explicates how microbial cells interact to enhance compressive strength [27]. Lower-grade concrete has less strength than higher-grade concrete, as calcite precipitation was greater in the latter after adding a concentration of 30–105 cfu/ml of *Bacillus megaterium* bacteria.

An increase by almost a quarter in strength in 50 MPa concrete—the highest grade of concrete—was ascertained to be the highest development rate of strength possible [10]. The incorporation of 10^5 cells/ml of *Sparcious pasteurii* bacteria occurred as well as fly ash concrete substituting 10% of the cement. The microorganism cell surface was covered by the deposition of calcium carbonate, which led to the structural fly ash concrete's strength improving by 20% [11]. The calcium carbonate precipitation enhanced the compressive strength of the silica fume concrete following the incorporation of the bacteria. The concrete contained calcium carbonate, as deduced following SEM and XRD microstructure analysis [28]. Over a period of 28 days, concrete not containing the *Sparcina pasteurii* bacteria was found to have 20% less compressive strength than concrete containing the bacteria [29].

Contrasting the control specimens, mortar compressive strength increased by 10%, 14%, and 19% via bacterial cells following distinct fly ash concentrations of 40%, 20%, and 10%, respectively substituting the cement [30]. Optimal efficacy in self-healing crack formations is an outcome of GNP that serves as an ideal carrier compound for the uniform distribution of bacteria. Irrespective of age, microbial precipitation of calcium carbonate led to improved compressive strength of concrete, upon the incorporation of GNP in tandem with *Bacillus subtilis* bacteria [3]. The addition of reactive spore powder within the cement mortar led to a notable improvement to compressive strength in comparison with the control cement mortar over the 28 days [5]. *Bacillus sp. CT-5* led to enhanced compressive strength in the mortar as the pores in the mortar and cement-sand matrix are filled in following the deposition of calcium carbonate on the surface of the cells [31–32]. The kind of calcium source given to the bacteria could alter the value of compressive strength, bacterial concentration, and the bacteria used, as illustrated in Table 6.2.

The core property that reflects concrete durability is permeability, as it regulates the entrance of deleterious materials into the concrete that cause deterioration under a pressure gradient. Micro-crack formations, tortuosity, connectivity, porosity, and size distribution make up the characteristics of the pore network within cementitious materials, also affecting permeability. The invasion of deleterious substances,

TABLE 6.2

Various Types of Bacteria and Their Compressive Strength Results

Bacteria Used	Best Results	Bacterial Concentration
Bacillus sp. CT-5	Compressive strength 40% more than the control concrete	5×107 cells/mm^3
Bacillus megaterium	Maximum rate of strength development was 24% achieved in highest grade of concrete 50 MPa	30×105 cfu/ml
Bacillus subtilis	Improvement of 12% in compressive strength as compared to controlled concrete specimens with light weight aggregates	2.8×108 cells/ml
Bacillus aerius	Increase in compressive strength by 11.8% in bacterial concrete compared to control 10% rice husk ash concrete	10^5 cells/ml
Sporosarcina pasteurii	Compressive strength 35% more than the control concrete	10^5 cells/ml

the particle size distribution, the water/cement (W:C) ratio, and the age of hardened cementitious materials regulate the mentioned features [33]. Concrete specimens experienced a reduction in permeability and water absorption following the deposition of calcium carbonate in concrete. A reduction in permeability and porosity in fly ash concrete was experienced following the incorporation of *S. pasteurii* bacteria, as per academic works [11]. Additionally, a 10^5 cells/ml concentration of bacteria in concrete led to a fourfold decrease in water absorption [28]. Microbial calcite deposition led to three times less water being absorbed by cubes cast with *Bacillus megaterium* and its nutrients in comparison with the control specimens [30]. The durability of structural concrete is enhanced via the calcite precipitation triggered by the incorporation of *Bacillus aerius*, which also leads to decreased porosity and water absorption [15]. Pores plugged in with calcium carbonate result in high to low permeability in AKKR5 bacteria concrete specimens with a concentration of 105 cells/ml, while a high to moderate permeability is demonstrated by all the control cement bag house filter dust concrete specimens, at the 28-day mark [34]. Water absorption of the recycled aggregate decreases following the microbial precipitation that enhances the caliber of the recycled aggregate [35–36].

The degradation of concrete structures occurs most frequently via chloride ingress which corrodes reinforcing steel. The internal pore structure of concrete determines the rate of chloride ingress into the material. Construction practices, the level of hydration, mix design, use of supplementary cementitious materials, and curing conditions are elements that determine the pore structure. Supervising a sample and how much electrical current passes through it is the method for enacting the rapid chloride permeability test. The permeability of a given kind of concrete receives a qualitative rating dependent upon the total charge that passed through the sample. The addition of bacteria in concreate can decrease the chances of chloride permeation in concrete. Concrete not containing bacteria was found to have on average 11.7% more coulombs than concrete containing bacteria. The mass reduction trend of sulfate exposed concrete was also found to be enhanced, on top of the decrease in chloride ingress in concrete when *Bacillus subtilis* and *Sparcious pasteurii* were added to the mix [29]. The charge that passes through the rice husk ash (RHA) and control concrete samples can be decreased through the incorporation of *Bacillus aerius* bacteria in concrete. The lowest charge passed at all curing ages is demonstrated by bacterial concrete. In comparison with standard concrete aged 7, 28, and 56 days, bacterial concrete samples experienced reductions of 55.8%, 49.9%, and 48.4%, respectively, in the total charge passed through them [34]. Rapid chloride penetration of 380 coulombs was resisted effectively when 10% silica fume concrete was combined with the best bacterial concentration of *Sparcious pasteurii* [28]. Despite a mere 762 coulombs penetration when the concrete contained 30% fly ash concrete, a drastic decrease in chloride ions was found in every fly ash concrete when 10^5 cells/ml concentration of *Sporoscarcina pasteurii* was added. The capability of concrete to repel the penetration of chloride ions determines the lifespan of concrete structures that interact with de-icing salts or marine environments.

SEM analysis enabled the visualization of calcite precipitation in concrete and mortar. Discovered within were rod-shaped bacteria that are linked with calcite crystals. The ingress of deleterious substances is blocked by the mentioned deposition that functions as an obstacle, resultantly reducing the permeability of the concrete [30]. Mineral precipitation leads to the enhancement of the concrete's microstructure following the incorporation of bacteria. XRD, EDS, and SEM analysis confirm this premise. In comparison with concrete without bacteria, and concrete with different volumes of added bacteria, a maximum weight of calcium, forming 38.76%, was achieved following the incorporation of 30×10^5 cfu/ml of *Bacillus megaterium*, as per an academic work [10]. Bacteria with distinct and diverse calcite crystals embedded within them were identified and visualized by SEM analysis. XRD and EDX analysis verified the presence of high quantities of calcium in the specimen, where the calcite took the form of calcium carbonate. Consequently, the durability of concreate is enhanced [36–37]. SEM images of the bacterial concrete and the control concrete are illustrated in Figure 6.3. The calcite crystals residing within the bacterial concrete can be verified by said SEM images [28]. SEM images also affirm that pores were filled in by the deposition of calcium carbonate following the incorporation of bacteria in RHA concrete which also enhanced the strength of the concrete. Bacterial RHA concrete and standard concrete are illustrated in the form of SEM images in Figure 6.3. Calcite evidently plugs in the gaps in the bacterial concrete, as seen in the images [34]. Microstructures led to findings being attained verifying that the crack formations in the test samples contained deposits of calcium carbonate within them. Hence, acid ingress, chloride permeability, and water absorption all diminished following the rise in signal transmission rate of ultrasonic pulse velocity [38].

6.4 SUMMARY

This chapter discussed bacteria as cementitious self-healing and their effects on concrete performance. Conclusions could be drawn as follows:

i. The importance of this chapter is to understand the use of urease-producing bacteria isolates, such as *Bacillus subtilis* and *Bacillus pasteurii* species, in healing of cracks in concrete.

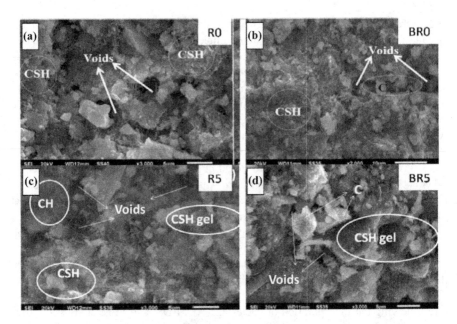

FIGURE 6.3 SEM images of: (a) normal concrete (R0); (b) bacterial concrete (BR0); (c) 5% of RHA Concrete (R5); and (d) bacterial concrete with 5% RHA (BR5) [1].

ii. The chapter has reviewed different types of bacteria that can be used as self-healing agents for cracks repair applications.

iii. Inclusion of bacteria as self-healing agent in the cementitious matrix positively enhances the compressive strength of proposed concrete.

iv. In term of durability advantage, the use of bacteria led to decreases in water penetration and chloride ion permeability.

v. The results of this chapter recommend that using "microbial concrete" can be an alternative and high-quality concrete sealant which is cost effective, environmentally friendly, and eventually leads to improvement in the durability of building materials.

REFERENCES

1. Vijay, K., M. Murmu, and S.V. Deo, *Bacteria based self healing concrete—A review.* Construction and Building Materials, 2017.**152**: p. 1008–1014.
2. Mostavi, E., et al., *Evaluation of self-healing mechanisms in concrete with double-walled sodium silicate microcapsules.* Journal of Materials in Civil Engineering, 2015.**27**(12): p. 04015035.
3. Khaliq, W., and M.B. Ehsan, *Crack healing in concrete using various bio influenced self-healing techniques.* Construction and Building Materials, 2016.**102**: p. 349–357.
4. Luo, M., C.-X. Qian, and R.-Y. Li, *Factors affecting crack repairing capacity of bacteria-based self-healing concrete.* Construction and Building Materials, 2015.**87**: p. 1–7.
5. Luo, M., and C. Qian, *Influences of bacteria-based self-healing agents on cementitious materials hydration kinetics and compressive strength.* Construction and Building Materials, 2016.**121**: p. 659–663.

6. Hung, C.-C., and Y.-F. Su, *Medium-term self-healing evaluation of engineered cementitious composites with varying amounts of fly ash and exposure durations.* Construction and Building Materials, 2016.**118**: p. 194–203.

7. Tziviloglou, E., et al., *Bacteria-based self-healing concrete to increase liquid tightness of cracks.* Construction and Building Materials, 2016.**122**: p. 118–125.

8. De Muynck, W., et al., *Bacterial carbonate precipitation improves the durability of cementitious materials.* Cement and Concrete Research, 2008.**38**(7): p. 1005–1014.

9. Siddique, R., et al., *Properties of bacterial rice husk ash concrete.* Construction and Building Materials, 2016.**121**: p. 112–119.

10. Andalib, R., et al., *Optimum concentration of Bacillus megaterium for strengthening structural concrete.* Construction and Building Materials, 2016.**118**: p. 180–193.

11. Chahal, N., R. Siddique, and A. Rajor, *Influence of bacteria on the compressive strength, water absorption and rapid chloride permeability of fly ash concrete.* Construction and Building Materials, 2012.**28**(1): p. 351–356.

12. Anne, S., et al., *Evidence of a bacterial carbonate coating on plaster samples subjected to the Calcite Bioconcept biomineralization technique.* Construction and Building Materials, 2010.**24**(6): p. 1036–1042.

13. Pei, R., et al., *Use of bacterial cell walls to improve the mechanical performance of concrete.* Cement and Concrete Composites, 2013.**39**: p. 122–130.

14. Jonkers, H.M., et al., *Application of bacteria as self-healing agent for the development of sustainable concrete.* Ecological Engineering, 2010.**36**(2): p. 230–235.

15. Siddique, R., and N.K. Chahal, *Effect of ureolytic bacteria on concrete properties.* Construction and Building Materials, 2011.**25**(10): p. 3791–3801.

16. Erşan, Y.Ç., et al., *Screening of bacteria and concrete compatible protection materials.* Construction and Building Materials, 2015.**88**: p. 196–203.

17. Dhami, N.K., M.S. Reddy, and A. Mukherjee, *Improvement in strength properties of ash bricks by bacterial calcite.* Ecological Engineering, 2012.**39**: p. 31–35.

18. Van Tittelboom, K., et al., *Use of bacteria to repair cracks in concrete.* Cement and Concrete Research, 2010.**40**(1): p. 157–166.

19. De Belie, N., and J. Wang, *Bacteria-based repair and self-healing of concrete.* Journal of Sustainable Cement-Based Materials, 2016.**5**(1–2): p. 35–56.

20. De Belie, N., *Application of bacteria in concrete: A critical evaluation of the current status.* RILEM Technical Letters, 2016(1): p. 56–61.

21. Wang, J., et al., *X-ray computed tomography proof of bacterial-based self-healing in concrete.* Cement and Concrete Composites, 2014.**53**: p. 289–304.

22. De Muynck, W., et al., *Bacterial carbonate precipitation as an alternative surface treatment for concrete.* Construction and Building Materials, 2008.**22**(5): p. 875–885.

23. Bollinger, J., et al., *Autonomic healing of polymer composites.* Letters to Nature, 2001.**409**: p. 1804–1807.

24. Souradeep, G., and H.W. Kua, *Encapsulation technology and techniques in self-healing concrete.* Journal of Materials in Civil Engineering, 2016.**28**(12): p. 04016165.

25. Ghosh, P., et al., *Use of microorganism to improve the strength of cement mortar.* Cement and Concrete Research, 2005.**35**(10): p. 1980–1983.

26. Zhang, Y., H. Guo, and X. Cheng, *Role of calcium sources in the strength and microstructure of microbial mortar.* Construction and Building Materials, 2015.**77**: p. 160–167.

27. Ramachandran, S.K., V. Ramakrishnan, and S.S. Bang, *Remediation of concrete using micro-organisms.* ACI Materials Journal-American Concrete Institute, 2001.**98**(1): p. 3–9.

28. Chahal, N., R. Siddique, and A. Rajor, *Influence of bacteria on the compressive strength, water absorption and rapid chloride permeability of concrete incorporating silica fume.* Construction and Building Materials, 2012.**37**: p. 645–651.

29. Nosouhian, F., D. Mostofinejad, and H. Hasheminejad, *Concrete durability improvement in a sulfate environment using bacteria.* Journal of Materials in Civil Engineering, 2016.**28**(1): p. 04015064.

30. Achal, V., X. Pan, and N. Özyurt, *Improved strength and durability of fly ash-amended concrete by microbial calcite precipitation.* Ecological Engineering, 2011.**37**(4): p. 554–559.

31. Achal, V., A. Mukherjee, and M.S. Reddy, *Microbial concrete: Way to enhance the durability of building structures.* Journal of Materials in Civil Engineering, 2011.**23**(6): p. 730–734.

32. Achal, V., et al., *Lactose mother liquor as an alternative nutrient source for microbial concrete production by Sporosarcina pasteurii.* Journal of Industrial Microbiology and Biotechnology, 2009.**36**(3): p. 433–438.

33. Phung, Q.T., et al., *Determination of water permeability of cementitious materials using a controlled constant flow method.* Construction and Building Materials, 2013.**47**: p. 1488–1496.

34. Siddique, R., et al., *Influence of bacteria on compressive strength and permeation properties of concrete made with cement baghouse filter dust.* Construction and Building Materials, 2016.**106**: p. 461–469.

35. Qiu, J., D.Q.S. Tng, and E.-H. Yang, *Surface treatment of recycled concrete aggregates through microbial carbonate precipitation.* Construction and Building Materials, 2014.**57**: p. 144–150.

36. Kim, H.-K., et al., *Microbially mediated calcium carbonate precipitation on normal and lightweight concrete.* Construction and Building Materials, 2013.**38**: p. 1073–1082.

37. Luo, M., and C.X. Qian, *Performance of two bacteria-based additives used for self-healing concrete.* Journal of Materials in Civil Engineering, 2016.**28**(12): p. 04016151.

38. Wiktor, V., and H.M. Jonkers, *Quantification of crack-healing in novel bacteria-based self-healing concrete.* Cement and Concrete Composites, 2011.**33**(7): p. 763–770.

7 Nanomaterials-Based Self-Healing Cementitious Materials

7.1 INTRODUCTION

Generally, self-healing is beneficial for the durability of materials. Particularly, it is advantageous when human interference is difficult, such as in construction purposes in the midst of harsh physical and chemical environments. Self-healing is also required to protect material characteristics, especially in kinetic and thermodynamic conditions that support large defect density like nanostructures. Nanomaterials invariably reveal excellent functional attributes. Compared to ordinary materials, nanomaterials degrade faster due to the presence of numerous interfacial atoms. Many functional nanostructures can be combined to fabricate diverse nanosystems, wherein some components can also be incorporated to offer self-healing actions. In fact, this strategy is simplistic compared to the design of a more robust nanosystem [1]. Of late, due to the rapid advancement of nanoscience and nanotechnology, the design and fabrication of self-healing materials has taken new frontiers, wherein materials with particle size below 500 nm are termed as nanomaterials. Self-healing materials can recover from damage autonomously. In many circumstances, the self-healing action can also be prompted via temperature as external stimuli; systems with this capacity are called non-autonomic self-healing materials [2].

7.2 SIGNIFICANCE OF NANOMATERIALS AS SELF-HEALING

The contribution of nanomaterials in enhancing the workability, strength, and durability of building materials can affect the hydration kinetics of cement [3]. Furthermore, the addition of nanomaterials can improve the performance of cement appreciably. The high performance of nanomaterials attracted researchers to apply nanomaterials to create sustainable concrete by self-healing technology. The use of Nano-Self-Healing concrete in construction industry would be associated with environmental costs. Careful analysis and selection of materials and how to combine them with cement can dramatically improve strength and durability, as well as greatly reduce the environmental impact of the life cycle. It was noted the nanomaterials can play an important role in the development of construction materials, and thus proper understanding of these nanomaterials is extremely important.

7.3 PRODUCTION OF NANOMATERIALS

It is well known that nanoparticles produce superior effect on filler to that of micron-sized materials. Guterrez [4] acknowledged that all materials can be converted

DOI: 10.1201/9781003195764-7

into nanoparticles via crushing or chemical treatment. The production accuracy of nanoparticles is decided by the purity and the chemical constituents of the parent materials. Two routes are followed for the large-scale production of nanomaterials: the top-down approach [5–6] and the bottom-up approach [7]. These approaches are chosen based on the appropriateness, price, and knowledge of the nanoscale properties of the material under consideration [8]. Milling techniques fall in the category of top-down approach, wherein the selection of milling machine is favored due to its accessibility, inexpensiveness, and feasibility of easy modification without requiring any chemical reagents or complex electronic equipment. In the top-down approach, big structures (bulk) are transformed to small ones (nanodimension) while keeping their actual physical or chemical behavior intact via atomic level control [9], which has been applied at industrial level. Using milling technique, a high volume of nanoparticles can be produced. Though the top-down approach is contemporary for nanofabrication, the consistency and superiority of the yields are often unpredictable. To overcome such drawbacks, the milling techniques (a top-down approach) can further be improved by increasing the number of balls, ball types, milling speed, and nature of the jar where the quality of nanoparticles become better [10].

High-energy ball milling techniques have been widely used to fabricate diverse nanomaterials such as nanoparticles, nanograins, nanoalloys, nanocomposites, and nanoquasicrystals. John Benjamin introduced this technique to produce oxide particles inside the matrix of nickel superalloys in 1970. Utilizing milling technique, the properties of alloy components effective for high thermal structure were altered and mechanical strengths were improved. Factors including fractures, plastic distortions/deformations, and cold welds during the milling process affect the materials' conversion into the desired morphology. Milling not only crushes the material into tinier fragments but also blends numerous particles or materials to transform them into new phases with different compositions. Usually, the final yields of milling technique are in the form of flakes where refinements are performed based on the selected ball and milling standard. Most of the nanomaterials (nanosilica, nanoalumina, nanoclay) utilized in concrete can be obtained via bottom-up approach, which is adopted for materials engineering at atomic or molecular levels via the self-assembling process—so-called molecular-nanotechnology or molecular-level processing. It is applied indirectly in nanomaterials and chemical production [11]. Varied morphology of nanoparticles obtained via bottom-up strategy is often customized through chemical synthesis technique. In comparison to top-down strategy, the bottom-up strategy can generate nanomorphology with more uniformity and reproducibility. Besides, using bottom-up approach, one can produce novel nanocrystals with perfect atomic or molecular ordering. The bottom-up approach based production of nanomaterials is useful for achieving high electronic conductivity, optical absorption, and chemical reactivity [12]. Through bottom-up strategy, one can get tinier size and formation of consistent surface atoms with modified surface energies and morphologies. Usually, the bottom-up technique is exploited to prepare self-healing and self-cleaning nanomaterials with improved catalytic properties, sensing capacities, and novel pigment characteristics. Moreover, the main drawbacks of the bottom-up scheme is related to high running cost, need of experts for chemical synthesis, and limitations of laboratory orientation only [13]. The nanoparticles produced using the bottom-up approach is excellent for advance applications, including electronic components and biotechnology.

7.4 NANOMATERIALS-BASED CONCRETES

In recent years, the growth in nanotechnology and the accessibility of nanomaterials suited for construction usage including nanosilica, nanoalumina, polycarboxylates and nanokaolin have improved the concrete properties remarkably [8, 14–15]. Intensive research has revealed that the mechanical properties such as compressive strength, splitting tensile, and flexural strength of cement pastes [16–17], mortars [14, 18], and concretes [19] can be improved via a tiny quantity of nanomaterials. Early strengths of pastes, mortars [14, 18], and concrete [19] in the presence of nanomaterials were reported to be much higher than those formulated with conventional OPC. Development of such higher strengths were ascribed to the faster cement hydration process and pozzolanic reaction, reduction of pores density, and enhanced interfacial bonding amid hardened cement paste and constituents (aggregates). Nanomaterials were also exploited to reduce the porosity and enhance the durability properties of concrete [20–21]. With developing the concrete technology, the nanomaterials were used in many applications (Figure 7.1) and the self-healing technology was one of the important application to produce sustainable and smart concrete.

7.5 PRODUCTION OF NANOCONCRETE

Using materials with a particle size less than 500 nm in concrete production as admixture or part cement replacement is called nanoconcrete. It was shown that the strength of normal concrete tends to be enhanced with the inclusion of nanoparticles. The bulk properties and packing model structure of concrete can remarkably be improved via the incorporation of nanoparticles. Nanoparticles act as excellent filling agents through the refinement of intersection zones in cementitious materials and production of high density concrete. The manipulation or modification of these nanoparticles in the cement matrix can render a new-fangled nanostructure [22–24]. General deficiencies in the microstructures of concretes including voids, micro-porosity, and corrosion originating from the reaction of alkaline silica can be discarded. The advancement of nanomaterials occurred due to their characteristics as new binding agents with particle sizes much tinier than traditional OPC. This

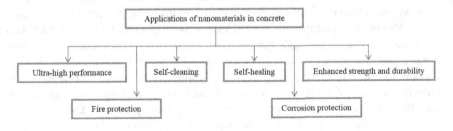

FIGURE 7.1 Applications of nanomaterials in concrete.

property enhances the hydration gel product by imparting a neat and solid structure. Also, by using a blend of filler and extra chemical reaction in the hydration scheme, high performing novel nanoconcrete with enhanced durability can be achieved.

The application of nanotechnology in concrete is still in its infancy. Ever-growing demand for ultra-high-performance concrete (UHPC) and recurring environmental pollution caused by OPC prompted engineers to exploit nanotechnology in construction materials. Classical blends of UHPC with incorporated silica fumes can achieve enhanced strength and high durability. However, limited accessibility and high-pricing of nanomaterials not only slowed down the growth of UHPC technology but also made it less demanding compared to conventional high-strength concrete (HSC). To overcome these limitations, nanotechnology and production of UHPC emerged in its own right wherein an alternative to silica fume has been developed. Exploiting the nanoproduction idea, a typical nanomaterial mimicking the attributes of silica fume was designed. Nanosilica is certainly the newest material in nanotechnology-based processing that has been used as a substitute for silica fume [25]. Using this celebrated nanosilica component, several types of nanoparticles have been synthesized which are effective for concrete production [26], nanoalumina [27], titanium oxide nanoparticles [28], carbon nanotube [29], and nanopolycarboxylates [30], and are the emergent nanomaterials in the new nanoconcrete era. It is now customary to discuss about the production and possible applications of nanomaterials.

7.6 NANOMATERIALS-BASED SELF-HEALING CONCRETE

Advancement of self-healing materials in concrete industries is an emerging trend. The generic term self-healing concrete is used for cement-based materials that repair themselves after the material or structure becomes damaged via diverse deterioration mechanisms. Nowadays, using nanomaterials for sustainable concrete also gets a lot of attention around the world. Combining the two technologies (self-healing and nanomaterials) together contributed to enhancing the durability of concrete and successfully led to producing sustainable concrete [31–32]. Most application of nanomaterials in self-healing has been used to control the corrosion of steel bars in reinforcement concrete. Koleva [31] has reported the ability to improve reinforced concrete performance by incorporating nanoscale materials with tailored properties—i.e., core-shell polymer vesicles or micelles—in a cement-based system. Little research has been conducted into use of nanomaterials for self-healing concrete.

Qian et al. [33] studied the curing situations in the presence of air, CO_2, and water in the wet state as well as dry state. Furthermore, the effects of nanoclay with water (used as inner water furnishing agent for hydration) on micro-cracks were determined. Results revealed that the healing extent could be considerably superior with the inclusion of nanoclay and better cementitious materials concentration for mixes. It was stated that for all studied air-cured mixtures, the final crack did not occur at the new location, indicating good healing. Despite the interior water supply from nanoclay, it revealed comparatively weak strengthening, enabling it difficult to relocate the final crack sites from pre-existing ones to other positions. This unproductive recovery with air curing was also reported by Qian et al. [33]. It was shown that the self-healing behavior of ECC with superabsorbent polymer (SAP) capsules with water as the interior pool for extra hydration was effective. Meanwhile, various repairing products were recognized across the cracked facades. However, noticeable healed cracks were absent completely. It was acknowledged that the cracks could very likely undergo the self-recovery mechanism without much effectiveness. The significance of enough water or moisture accessibility was highlighted that acted as the reactant for additional hydration, as well as the calcium, aluminum, and silica ions transportation media.

7.7 NANOSILICA

In sustainable concrete development, nanosilica (SiO_2) is a gifted nanomaterial broadly utilized in UHPC. In general, nanosilica (Figure 7.2) is manufactured from micron-sized silica. Strong reactions produced by nanosilica in UHPC are comparable to silica fume or micro-silica involving their strength, performance, and durability improvement [34–36]. Qing et al. [35] demonstrated that concrete with nanosilica can gain strength earlier than that of silica fume. It was shown that the incorporation of nanosilica in concrete could enhance its workability when the inclusion of super-plasticizers amount was lowest. Besides, nanosilica particles' size could act as ultra-filler inside the concrete, leading to dense and refined micro-voids to impart smart microstructures [37]. Another benefit involving nanosilica is better control of the water-to-cement ratio, wherein the strength can be customized easily. Quercia et al. [38] reported that the incorporation of nanosilica at certain dosage improved the strength of concrete and performed as cement substitution component. Nearly 20–30% of cement could be lowered via the implementation of nanosilica; thereby, it can be a good substitute for cement. In self-healing applications, the capacity of nanosilica to react with available $Ca(OH)_2$ in concrete and formulate (C-S-H gel) was examined [39–42]. Using nanosilica, the self-healing concrete technology as additional mineral admixtures or in nanoencapsulation method has been studied. However, the disadvantage of nanosilica is the cost and availability in certain nations where nanosilica needs to be imported for use in the concrete industry [11].

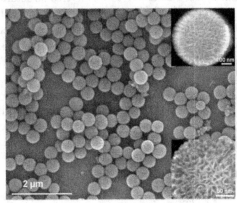 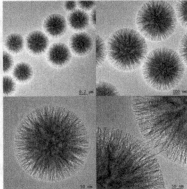

FIGURE 7.2 FSSEM nanosilica image [43].

7.8 NANOALUMINA

In cement hydration, it is known that the silica and alumina are two main components involved to formulate C-(A)-S-H gels with calcium. The role of silica inside cement was to alter its strength, whereas alumina controlled the setting time. Nanoalumina is produced from alumina, whereas the use of nanoalumina in concrete is seldom reported. The inclusion of nanoalumina in concrete—particularly high-performance concrete—can enormously affect concrete characteristics because it regulates the setting time of cement [44–45]. Presence of nanoalumina within cement can accelerate the early setting time for high-performance concrete, which in turn diminishes segregation and flocculation. In high-performance concrete mixes, any disturbance in the cement creates inhomogeneity and affects the working performance. Nanoalumina in high-performance concrete functions as a dispersive mediator in cement particulates [46–47]. Besides, nanoalumina can refine the porosity of hydration-gel product as nanofiller because the cement proportion in high performing concrete is high. Therefore, grains size distribution in such concrete is essential simultaneous with silica-mediated hydration. Without nanoalumina-mediated refinement, the hydration mechanism is weaker since the silica component cannot penetrate the interior structure of the gel. Incorporation of nanoalumina can create a path for easy injection of the silica or binding materials into the interior microstructures of the hydration gel to start the refinement [48–49]. All these merits make nanoalumina one of the important materials that can be used in the future to produce smart and sustainable concrete by using self-healing and nanotechnology.

7.9 CARBON NANOTUBE

Carbon nanotube (CNT) is a carbon allotrope having a cylindrical nanostructure. CNTs can be produced with length-to-diameter (aspect) ratio of up to 132,000,000:1, considerably more than any other material. They possess atypical properties, which are precious for applications in such diverse fields as materials science and nanotechnology [4, 11]. Particularly, due to their unusual thermal conductivity, mechanical, and electrical characteristics, CNTs can be used as additives for diverse structural applications. NTs are affiliated with the fullerene family, in which long and hollow construction with walls created by one-atom-thick sheets of carbon—so-called grapheme—makes them different. Such sheets can be rolled at definite orientations, whereby the revolving angle and radius together determine the attributes of CNTs [29]. Nanotubes are divided into single-walled NT (SWNT) and multi-walled NT (MWNT). Individual NTs are aligned in the form of "ropes" that hold them mutually by van der Waals forces or via pi-stacking [50–51]. These ropes make chemical bonds in the CNT structure, whereby the bonds of NTs are comprised alike the graphite. These bonds are much stronger than the ones formed in alkanes and diamond, making CNTs exceptionally strong.

Flexibility is one of the significant merits of CNT for production of sustainable concrete. Using CNT, the sustainable concrete design may be changed into distinctive or rigid types. The flexible nature of CNT is advantageous for enhancing the sustainable concrete strength. Compared to other nanomaterials, CNT is superior in terms of flexibility improvement and strength enhancement of sustainable concrete [29]. Predominantly, the dimension of CNT is tinier than other nanomaterials. The major function of CNT in sustainable concrete is to advance the stress and compressive strength [29]. CNT can be used in self-healing concrete using the engineered cementitious composite (ECC) method and will contribute to produce sustainable and smart concrete for the construction industry in the future. Nevertheless, deficiency of CNT resources may trim down the attention toward potential concrete application of this material.

7.10 POLYCARBOXYLATES

In the past, polycarboxylates (PCE) nanomaterials have been utilized in concrete [52]. PCE is a polymer-based compound that is obtained from methoxy-polyethylene glycol co-polymer. It functions in the secondary or side reactions which are reinforced to methacrylic acid co-polymer as the major element. Usually, the carboxylate group is comprised of a water molecule, rendering a negative charge alongside the backbone of PCE. The polyethylene oxide group provides a non-uniform electron cloud distribution and chemical polarities to the secondary or side reactions. The number and the length of secondary or side groups are can easily be changed. In case the secondary or side reactions possess numerous electrons, it lowers the large molar mass and alters the polymer density, resulting in reduced performance of cement suspension [11, 53]. For both chains to combine and pair simultaneously, longer side groups and strong charge density from one to another reaction end must form. Usually, polycarboxylate is used in concrete as high range water reducer (HRWR). Inclusion of PCE allows in controlling the concrete workability better at lower water-to-cement ratios.

The characteristics of PCE in concrete rely on its amount, where elevated amounts may cause false setting without any hydration incidence in the concrete [54]. As an aside, the addition of PCE at required levels can create self-compacting concrete (SCC) which enhances the workability of concrete. It also creates a flow concrete with enormous effects at small and elevated intensity regions. Other benefits of PCE inclusion in UHPC or concrete is related to its capacity to be applied in marine atmospheres. Pores or voids in UHPC can remarkably be reduced to get more compact structure in the presence of PCE because it can eliminate air bubbles and improve the concrete density. Furthermore, refinement of UHPC microstructure can avoid or reduce the permeability rate under marine conditions, lowering the attack from seawater containing sulfate and chloride ions. On top of this, the use of polycarboxylate is regarded as relatively more green strategy than the use of silica fume and other stabilizers in the UHPC microstructures refinement.

Birgisson et al. [55] used PCE in HSC instead of silica fume to improve the workability and performance of UHPC. Additionally, approximately 70% of constituent materials in UHPC—including silica fume, superplasticizer, and fiber—were reduced. Ultimately, it was shown that the incorporation of PCE within UHPC could improve the overall performance compared to pure UHPC or HSC. It was acknowledged that PCE addition of 2.5% of cement weight in HSC could rapidly enhance the early-age strength. Consequently, on the first day the HSC strength was enhanced from 40 to 80 MPa. Meanwhile, at 28 days, obtained strength was about 70–100 MPa at low PCE level, which proved that PCE can act as a good substitute to improve the concrete performance. In short, simple management with the fewest parameters or protocols of PCE makes it celebrated nanomaterials to be incorporated in UHPC.

7.11 TITANIUM OXIDE

Titania or titanium oxide (TiO_2) has widely been used as pigment in food coloring, paints, photocatalyzes, implants, solar cells, and many other applications. Normally, it is resourced from ilmenite, rutile, and anatase phases, and exists in nature as mineral phases such as rutile, anatase and brookite. At high pressure, TiO_2 is transformed to monoclinic baddeleyite and orthorhombic structures, recently discovered at the Ries crater in Bavaria [50, 56]. Ilmenite and rutile are the most abundant forms of ore containing titanium dioxide worldwide (98%). Heating above temperatures 600–800°C, one achieves the metastable anatase and brookite phases [57].

Incorporation of TiO_2 into UHPC and other concretes revealed remarkable influence on self-cleaning capability and contributed as green material implementation in engineering construction [56]. Jubilee Church in Rome (Italy) exploited this self-cleaning attribute of TiO_2 in buildings, materials, paving, and product finishing [8]. TiO_2 provides accelerated strength of concrete at early age, wherein the concrete's performance and abrasion resistance both are enhanced [58]. TiO_2 acts as a glassy layer (extreme porosity) or pigment in the exterior of the particles in UHPC; concrete modifies their microstructure—and thus, the performance. Moreover, such layers can react with the hydration gel products in mixing by producing protective layers, thereby imparting self-cleaning capacity to the surface of concrete. This self-cleaning action of titania in the exterior and coated concrete surface make them extremely durable with permeability. This enhanced activity of titania-incorporated concrete is regarded as a fiber-reinforced system that mimics the glassy fiber effect. It is realized that refinement and tailoring of the hydration gel via such fibers can be useful for the production of concrete with improved strength and prolonged durability [59]. Yet, health safety is a major concern related to the usage of TiO_2 because of its dusty nature, whereby small dosage can result in notable environmental impact, especially to workers during packaging and manufacturing. Some reports exhibited that TiO_2 generates inflammation and causes cancer in factory workers [58]. Consequently, it must be handled with care, particularly during mixing and processing of TiO_2.

7.12 NANOKAOLIN

Nanokaolin is a byproduct of kaolin also called kaolinite, a very important industrial clay mineral with the chemical formula of $Al_2Si_2O_5(OH)_4$ [60–61]. This is a silicate mineral with many layers, whereby one tetrahedral layer is connected via oxygen atoms to another octahedral layer of alumina [62]. Kaolinite-enriched rocks are generally called kaolin or China clay [63]. This clay encloses a white mineral called dioctahedral phyllosilicate that is formed by chemical changes of aluminum silicate minerals including feldspar [64]. For ceramic applications, kaolin is heat treated and altered into $Al_2O_3.2SiO_2.2H_2O$. After treatment or endothermic dehydration crystalline phase, kaolin is transformed to amorphous structures [65]; the new phase is known as metakaolin. It contains amorphous silica and alumina with some hexagonal layering [66], which is very reactive pozzolan and reveals comparable reaction as silica fume. Furthermore, by refining the microstructure of metakaolin, both strength can durability can be improved, thereby allowing consistent water penetration. These properties make kaolin more strong and cost-effective than silica fume [11, 67].

Nanokaolin is generated following either top-down or bottom-up approach, whereby the final product is influenced by the processing conditions. Usually, the generation of nanokaolin is comprised of layering or stacking flakes. Apparently, the kaolin particles are identical to the nanokaolin, where the morphologies after the size conversion from microparticles to nanoparticles offer wider surface area. In the concrete, nanokaolin is treated to form an extra reactive or stable product called nanometakaolin. Being a newly used supplementary ingredient in concrete, the presence of nanometakaolin improves the concrete properties unexpectedly [11]. The positive effects of metakaolin in UHPC and other kinds of concretes are demonstrated [21]. Morsy et al. [68] showed that nanometakaolin inclusion in concrete could improve mortar compressive strength by about 8–10%. Interestingly, the tensile and flexural strength enhancement of mortar containing nanometakaolin was discerned to be nearly 10–15% compared to plain OPC [68–69]. Despite the advantages of nanometakaolin incorporation into mortars to improving the performance, some shortcomings on the application of in UHPC were identified. Actually, lack of raw kaolin in certain nations makes nanometakaolin less popular than celebrated silica fume. Thus, standard guidelines and commercial protocols must be developed for large-scale production of nanokaolin and nanometakaolin as useful nanomaterial alternatives to concrete for serving the construction industry.

7.13 NANOCLAY

Nanoparticles of layered mineral silicates are called nanoclay. Based on these nanoparticles, chemical compositions, and morphologies, nanoclays are categorized into many classes including montmorillonite, bentonite, kaolinite, hectorite, and halloysite. It is one of the most inexpensive materials with beneficial outcomes in polymers. It is prepared from deposits of montmorillonite mineral, having platelet structures of average thickness of 1 nm thick and width of 70–150 nm. It possesses many good qualities and is an outstanding base material for nanotechnology maneuvering. These notable attributes are the stability, interlayer spacing, elevated hydration, and swelling capacity, as well as strong chemical reactivity. Clays and their improved organic products can be analyzed via simple and modern instruments that can evaluate the chemical compositions. These tools include gravimetric analysis, inductively coupled plasma (ICP) or X-ray fluorescence (XRF) spectroscopy, cation exchange capacity (CEC) using standard ammonium acetate technique, surface area determination, Fourier transform infrared spectroscopy (FTIR), powdered X-ray diffraction (PXRD), and so on [58, 70–72]. These clays are also distinguished via cation exchange capacity that can differ broadly based on the source and nature of clay. The clay purity can influence the nanocomposite behaviors. Therefore, it is significant to obtain montmorillonite with minimal impurities such as crystalline silica (quartz), amorphous silica, calcite, and kaolin [70]. For the purification of clay, several techniques have been developed such as hydrocyclone, centrifugation, sedimentation technique, and chemical treatment [66].

Clays are regarded as economical and easily accessible materials. Despite their abundance worldwide, the guideline and methods for transforming clay to nanomaterials is not yet well documented. It is thus important to explore the constructional benefits and disadvantages of nanoclay as potential materials, although it has been diversely employed in the polymeric system. Nonetheless, the proof of the improvement in the material hardness, thermal stability, barrier coating, and solvents, together with the enhancement in the electronic and novel types of materials, are required. In the construction process, nanoclay is used as an additive to improve the mechanical and binding concrete characteristics. Morsy et al. [68] acknowledged that the enhancement in the compressive and tensile strength of mortar cement in the presence of nanoclay as additive. Furthermore, the thermal properties of concretes can be enhanced via the inclusion of nanoclay as cement additive in the paste [71–72]. Qian et al. [33] studied the inclusion of nanoclay with water (worked as the inner water provider to promote hydration alongside the micro-cracks). It was shown that the recovery level could be considerably improved with the incorporation of nanoclay in the mixtures. The mechanism of healing depended on the reaction between calcium hydroxide and nanoclay to formulate C-S-H gel leading to healing the crack.

7.14 NANOIRON

Copper, cobalt, and nickel are ferromagnetic materials with very limited applications due to their toxicity and susceptibility to oxidation. Unlike these, iron oxide nanoparticles are attractive, owing to their super paramagnetic properties and their potential applications in many fields. They have many applications in construction industry, but of particular interest is as a coloring and anti-corrosion agent in construction materials and coatings. Iron oxide nanoparticles have very good UV blocking capabilities, making these nanoparticles ideal for glass applications ranging from glass coatings to sunglasses. They also allow for better dispersion in paints and coatings, especially in high gloss and automotive applications [73–74]. The possibility and quick reaction between $Ca(OH)_2$ and Fe_2O_3 nanoparticles led to formation of high amounts of reaction products to close the cracks.

7.15 NANOSILVER

Nanosilver is now well known for anti-bacterial, anti-viral and anti-fungal efficacy, as well as for cellular metabolism in inhibiting growth of cells. Nanosilver can slow down growth multiplication of bacterial and fungal infections that cause bad odor, itching, and sores. Using nanotechnology tools, it is possible now to produce silver nanoparticles of accurate morphology (size and shape) with very uniform distribution. Surface coating of nanoparticles from diverse materials can enhance the surface area by several orders of magnitude compared to their bulk counterpart. Particularly, silver is attractive metal because of its extraordinary size- and shape-dependent optical characteristics, efficient plasmon excitation, and high electrical and thermal conductivity in the bulk among all metals. These unique features allow silver nanoparticles (Ag NPs) for potential applications in catalysis, selective oxidation of styrene, antimicrobial coatings, optical sensing, printed electronics, and photonics. Ag NPs of dimension in the range of 1–100 nm find several applications in the field of catalytic processes, photonic devices, electronics, optoelectronic, etc., due to their distinct physical, structural, morphological, chemical, electrical, optical, and magnetic attributes. On top of this, Ag NPs have widely been used as antibacterial and antifungal agents in the biomedical field, in textile engineering, for water treatment, and for several other consumer products [73–74].

7.16 SUMMARY

In recent times, production of sustainable concrete via self-healing technology became useful in construction industries worldwide. Exponential increase in the usage of OPC caused severe environmental damage. The immense benefits and usefulness of self-healing concrete technology were demonstrated in terms of its sustainability, energy saving traits, and environmental affability. The foremost challenges, current progress, and future trends of nanotechnology related to self-healing concrete's potential were emphasized. An all-inclusive overview of the appropriate literature on nanomaterials-based self-healing concrete allowed us to draw the following conclusions:

i. Self-healing concretes are characterized by many significant traits such as being less polluting, cheap, and eco-friendly, and demonstrating elevated durability performance in harsh environments. These properties make them effective sustainable materials in the construction industry.
ii. The design of nanomaterials-based self-healing concretes with improved performance and endurance useful for several applications is a new avenue in nanoscience and nanotechnology.
iii. Environmental pollution can considerably be reduced by implementing high strength and durable cementitious composites fabricated using diverse nanoparticles, carbon nanotubes, and nanofibers.
iv. In the domain of building and construction, the production of materials via nanotechnology is going to play a vital role toward sustainable development in the near future.
v. Use of nanomaterials in concrete is advantageous in terms of improved engineering properties of cementitious materials, especially for the generation of self-healing and sustainable concretes.
vi. This comprehensive review is believed to provide taxonomy to navigate and underscore research progress toward nanomaterials-based self-healing concrete technology.

REFERENCES

1. Gupta, S., S. Dai Pang, and H.W. Kua, *Autonomous healing in concrete by bio-based healing agents—A review*. Construction and Building Materials, 2017.**146**: p. 419–428.

2. Wang, J., et al., *Use of silica gel or polyurethane immobilized bacteria for self-healing concrete*. Construction and Building Materials, 2012.**26**(1): p. 532–540.

3. Huseien, G.F., et al., *Geopolymer mortars as sustainable repair material: A comprehensive review*. Renewable and Sustainable Energy Reviews, 2017.**80**: p. 54–74.

4. Guterrez, K., *How nanotechnology can change the concrete world*. American Ceramic Society Bulletin, 2005.**84**(11): p. 1–5.

5. Abdoli, H., et al., *Effect of high energy ball milling on compressibility of nanostructured composite powder*. Powder Metallurgy, 2011.**54**(1): p. 24–29.

6. Shah, K.W., and G.F. Huseien, *Biomimetic self-healing cementitious construction materials for smart Buildings*. Biomimetics, 2020.**5**(4): p. 47.

7. Jankowska, E., and W. Zatorski. *Emission of nanosize particles in the process of nanoclay blending*. 3rd International conference on quantum, nano and micro technologies, 2009. ICQNM'09, IEEE.

8. Sanchez, F., and K. Sobolev, *Nanotechnology in concrete—A review*. Construction and Building Materials, 2010.**24**(11): p. 2060–2071.

9. Shah, S.P., et al., Nanoscale modification of cementitious materials. In *Nanotechnology in construction 3*. 2009: Springer, Netherlands. p. 125–130.

10. Saleh, N.J., R.I. Ibrahim, and A.D. Salman, *Characterization of nano-silica prepared from local silica sand and its application in cement mortar using optimization technique*. Advanced Powder Technology, 2015.**26**(4): p. 1123–1133.

11. Norhasri, M.M., M. Hamidah, and A.M. Fadzil, *Applications of using nano material in concrete: A review*. Construction and Building Materials, 2017.**133**: p. 91–97.

12. Gesoglu, M., et al., *Properties of low binder ultra-high performance cementitious composites: Comparison of nanosilica and microsilica*. Construction and Building Materials, 2016.**102**: p. 706–713.

13. Paul, K.T., et al., *Preparation and characterization of nano structured materials from fly ash: a waste from thermal power stations, by high energy ball milling*. Nanoscale Research Letters, 2007.**2**(8): p. 397.

14. Li, H., et al., *Microstructure of cement mortar with nano-particles*. Composites Part B: Engineering, 2004.**35**(2): p. 185–189.

15. Shah, K.W., et al., *Aqueous route to facile, efficient and functional silica coating of metal nanoparticles at room temperature*. Nanoscale, 2014.**6**(19): p. 11273–11281.

16. Porro, A., et al., *Effects of nanosilica additions on cement pastes*. Applications of nanotechnology in concrete design: Proceedings of the international conference held at the University of Dundee, Scotland, UK on 7 July 2005, Thomas Telford Publishing.

17. Qing, Y., et al., *Influence of nano-SiO 2 addition on properties of hardened cement paste as compared with silica fume*. Construction and Building Materials, 2007.**21**(3): p. 539–545.

18. Jo, B.-W., et al., *Characteristics of cement mortar with nano-SiO$_2$ particles*. Construction and Building Materials, 2007.**21**(6): p. 1351–1355.

19. Schoepfer, J., and A. Maji, *An investigation into the effect of silicon dioxide particle size on the strength of concrete*. Special Publication, 2009.**267**: p. 45–58.

20. Said, A.M., and M.S. Zeidan, *Enhancing the reactivity of normal and fly ash concrete using colloidal nano-silica*. Special Publication, 2009.**267**: p. 75–86.

21. Zhang, M.-H., and J. Islam, *Use of nano-silica to reduce setting time and increase early strength of concretes with high volumes of fly ash or slag*. Construction and Building Materials, 2012.**29**: p. 573–580.

22. Aydın, A.C., V.J. Nasl, and T. Kotan, *The synergic influence of nano-silica and carbon nano tube on self-compacting concrete*. Journal of Building Engineering, 2018.**20**: p. 467–475.

23. Lim, N.H.A.S., et al., *Microstructure and strength properties of mortar containing waste ceramic nanoparticles*. Arabian Journal for Science and Engineering, 2018: p. 1–9.

24. Fu, J., et al., *Comparison of mechanical properties of CSH and portlandite between nano-indentation experiments and a modelling approach using various simulation techniques.* Composites Part B: Engineering, 2018.**151**: p. 127–138.

25. Yu, R., P. Spiesz, and H. Brouwers, *Effect of nano-silica on the hydration and microstructure development of ultra-high performance concrete (UHPC) with a low binder amount.* Construction and Building Materials, 2014.**65**: p. 140–150.

26. Adak, D., M. Sarkar, and S. Mandal, *Effect of nano-silica on strength and durability of fly ash based geopolymer mortar.* Construction and Building Materials, 2014.**70**: p. 453–459.

27. Silva, J., et al., *Influence of nano-SiO2 and nano-Al2O3 additions on the shear strength and the bending moment capacity of RC beams.* Construction and Building Materials, 2016.**123**: p. 35–46.

28. Massa, M.A., et al., *Synthesis of new antibacterial composite coating for titanium based on highly ordered nanoporous silica and silver nanoparticles.* Materials Science and Engineering: C, 2014.**45**: p. 146–153.

29. Morsy, M., S. Alsayed, and M. Aqel, *Hybrid effect of carbon nanotube and nano-clay on physico-mechanical properties of cement mortar.* Construction and Building Materials, 2011.**25**(1): p. 145–149.

30. Navarro-Blasco, I., et al., *Assessment of the interaction of polycarboxylate superplasticizers in hydrated lime pastes modified with nanosilica or metakaolin as pozzolanic reactives.* Construction and Building Materials, 2014.**73**: p. 1–12.

31. Koleva, D., *Nano-materials with tailored properties for self healing of corrosion damages in reinforced concrete, IOP self healing materials.* 2008: SenterNovem, The Netherlands.

32. Gajanan, K., and S. Tijare, *Applications of nanomaterials.* Materials Today: Proceedings, 2018.**5**(1): p. 1093–1096.

33. Qian, S., J. Zhou, and E. Schlangen, *Influence of curing condition and precracking time on the self-healing behavior of engineered cementitious composites.* Cement and Concrete Composites, 2010.**32**(9): p. 686–693.

34. Indumathi, P., S. Shabhudeen, and C. Saraswathy, *Synthesis and characterization of nano silica from the Pods of Delonix Regia ash.* International Journal of Advanced Engineering Technology, 2011.**2**(4): p. 421–426.

35. Qing, Y., et al., *Influence of nano-SiO$_2$ addition on properties of hardened cement paste as compared with silica fume.* Construction and Building Materials, 2007.**21**(3): p. 539–545.

36. Bai, P., et al., *A facile route to preparation of high purity nanoporous silica from acid-leached residue of serpentine.* Journal of Nanoscience and Nanotechnology, 2014.**14**(9): p. 6915–6922.

37. Lindgreen, H., et al., *Microstructure engineering of Portland cement pastes and mortars through addition of ultrafine layer silicates.* Cement and Concrete Composites, 2008.**30**(8): p. 686–699.

38. Quercia, G., G. Hüsken, and H. Brouwers, *Water demand of amorphous nano silica and its impact on the workability of cement paste.* Cement and Concrete Research, 2012.**42**(2): p. 344–357.

39. Raki, L., et al., *Cement and concrete nanoscience and nanotechnology.* Materials, 2010.**3**(2): p. 918–942.

40. Lopez-Calvo, H., et al., *Compressive strength of HPC containing CNI and fly ash after long-term exposure to a marine environment.* Cement and Concrete Composites, 2012.**34**(1): p. 110–118.

41. Grinys, A., V. Bocullo, and A. Gumuliauskas, *Research of alkali silica reaction in concrete with active mineral additives.* Journal of Sustainable Architecture and Civil Engineering, 2014.**6**(1): p. 34–41.

42. Wang, L., et al., *Effect of nano-SiO2 on the hydration and microstructure of Portland cement.* Nanomaterials, 2016.**6**(12): p. 241.
43. Polshettiwar, V., et al., *High-surface-area silica nanospheres (KCC-1) with a fibrous morphology.* Angewandte Chemie International Edition, 2010.**49**(50): p. 9652–9656.
44. Wu, H., et al., *Modification of properties of reinforced concrete through nanoalumina electrokinetic treatment.* Construction and Building Materials, 2016.**126**: p. 857–867.
45. Sikora, P., M. Abd Elrahman, and D. Stephan, *The influence of nanomaterials on the thermal resistance of cement-based composites—A review.* Nanomaterials, 2018.**8**(7): p. 465.
46. Nazari, A., and S. Riahi, *Improvement compressive strength of concrete in different curing media by Al2O3 nanoparticles.* Materials Science and Engineering: A, 2011.**528**(3): p. 1183–1191.
47. Hosseini, P., et al., *Effect of nano-particles and aminosilane interaction on the performances of cement-based composites: An experimental study.* Construction and Building Materials, 2014.**66**: p. 113–124.
48. Rosenqvist, J., *Surface chemistry of Al and Si (hydr) oxides, with emphasis on nano-sized gibbsite (α-Al (OH) 3).* 2002: Umea University, Sweden. p. 1–73.
49. Richardson, I., *The nature of CSH in hardened cements.* Cement and Concrete Research, 1999.**29**(8): p. 1131–1147.
50. Sobolev, K., et al., *Nanomaterials and nanotechnology for high-performance cement composites.* Proceedings of ACI Session on Nanotechnology of Concrete: Recent Developments and Future Perspectives, 2006: p. 91–118.
51. Lai, F., M. Zain, and M. Jamil. *Nano cement additives (NCA) development for OPC strength enhancer and Carbon Neutral cement production.* Proceedings the 35th conference on our world in concrete and structures, Singapore, 25–27 August 2010.
52. Piqué, T.M., H. Balzamo, and A. Vázquez, Evaluation of the hydration of Portland cement modified with polyvinyl alcohol and nano clay. In *Key engineering materials.* 2011: Trans Tech Publ, Switzerland, **466**. p. 47–56.
53. Shamsaei, E., et al., *Graphene-based nanosheets for stronger and more durable concrete: A review.* Construction and Building Materials, 2018.**183**: p. 642–660.
54. Helmy, S., and S. Hoa, *Tensile fatigue behavior of tapered glass fiber reinforced epoxy composites containing nanoclay.* Composites Science and Technology, 2014.**102**: p. 10–19.
55. Birgisson, B., et al., *Konstanin Sobolev, Nanotechnology in concrete materials.* Transportation Research Circular, Sciences Engineering Medicine, Washington, DC. Number E-C170, 2012.
56. Maravelaki-Kalaitzaki, P., et al., *Physico-chemical and mechanical characterization of hydraulic mortars containing nano-titania for restoration applications.* Cement and Concrete Composites, 2013.**36**: p. 33–41.
57. Vallee, F. *Cementitious materials for self-cleaning and de-polluting facade surfaces.* RILEM international symposium on environment-conscious materials and systems for sustainable development, RILEM Publications SARL, 2004.
58. Pacheco-Torgal, F., and S. Jalali, *Nanotechnology: Advantages and drawbacks in the field of construction and building materials.* Construction and Building Materials, 2011.**25**(2): p. 582–590.
59. Chen, J., and C.-S. Poon, *Photocatalytic construction and building materials: From fundamentals to applications.* Building and Environment, 2009.**44**(9): p. 1899–1906.
60. Liu, Q., Y. Zhang, and H. Xu, *Properties of vulcanized rubber nanocomposites filled with nanokaolin and precipitated silica.* Applied Clay Science, 2008.**42**(1–2): p. 232–237.
61. Bessa, M.J., et al., *Moving into advanced nanomaterials. Toxicity of rutile TiO2 nanoparticles immobilized in nanokaolin nanocomposites on HepG2 cell line.* Toxicology and Applied Pharmacology, 2017.**316**: p. 114–122.

62. Adamis, Z., and R.B. Williams, *Bentonite, kaolin and selected clay minerals*. 2005: World Health Organization, Geneva.
63. Sabir, B., S. Wild, and J. Bai, *Metakaolin and calcined clays as pozzolans for concrete: a review*. Cement and Concrete Composites, 2001.**23**(6): p. 441–454.
64. Chakchouk, A., et al., *Formulation of blended cement: Effect of process variables on clay pozzolanic activity*. Construction and Building Materials, 2009.**23**(3): p. 1365–1373.
65. Blanchart, P., S. Deniel, and N. Tessier-Doyen, Clay Structural Transformations during Firing. In *Advances in science and technology*. 2010: Trans Tech Publ, Switzerland, **68**: p. 31–37.
66. Zhang, D., et al., *Synthesis of clay minerals*. Applied Clay Science, 2010.**50**(1): p. 1–11.
67. Ghafari, E., H. Costa, and E. Júlio, *Critical review on eco-efficient ultra high performance concrete enhanced with nano-materials*. Construction and Building Materials, 2015.**101**: p. 201–208.
68. Morsy, M., S. Alsayed, and M. Aqel, *Effect of nano-clay on mechanical properties and microstructure of ordinary Portland cement mortar*. International Journal of Civil & Environmental Engineering IJCEE-IJENS, 2010.**10**(01): p. 23–27.
69. Morsy, M., et al., *Behavior of blended cement mortars containing nano-metakaolin at elevated temperatures*. Construction and Building Materials, 2012.**35**: p. 900–905.
70. Zbik, M., and R.S.C. Smart, *Nanomorphology of kaolinites: Comparative SEM and AFM studies*. Clays and Clay Minerals, 1998.**46**(2): p. 153–160.
71. Al-Salami, A., H. Shoukry, and M. Morsy, *Thermo-mechanical characteristics of blended white cement pastes containing ultrafine nano clays*. International Journal of Green Nanotechnology, 2012.**4**(4): p. 516–527.
72. Heikal, M., et al., *Behavior of composite cement pastes containing silica nano-particles at elevated temperature*. Construction and Building Materials, 2014.**70**: p. 339–350.
73. Rattan, A., P. Sachdeva, and A. Chaudhary, *Use of Nanomaterials in Concrete*. International Journal of Latest Research in Engineering and Technology, 2016.**2**(5): p. 81–84.
74. Olar, R., *Nanomaterials and nanotechnologies for civil engineering*. Buletinul Institutului Politehnic din Iasi. Sectia Constructii, Arhitectura, 2011.**57**(4): p. 109.

8 Factors Affecting Concrete Self-Healing Performance

8.1 IMPACT OF CRYSTALLINE ADMIXTURES ON SELF-HEALING

In construction, one of the materials most widely used is concrete. The formation of cracks occurs in concrete structures. Deleterious substances enter through such cracks and diminish the toughness of the cement. Corrosion becomes more likely as the concrete's durability diminishes from such cracks [1–2]. The incorporation of autonomously healing materials into the concrete mix can counteract such issues. Crack formation width will be diminished while concrete toughness improves via the addition of such materials. The restoration of the strengths of the cracked materials is a characteristic of the cementitious material that forms part of autonomously healing concrete [3]. Microencapsulation, regular encapsulation, vascular, and direct methods are among the approaches for incorporating healing agents into the concrete [4]. The adoption of artificial techniques enhances the efficacy of autonomous healing in concrete, as per numerous investigations. Improved outcomes are observed when water immersion conditions are combined with the incorporation of crystalline admixtures in concrete. The flexural strength of concrete was enhanced in many studies by the incorporation of fibers and polymers with greater proportion ratios.

Self-healing concrete is the name given to concrete that contains autonomous healing substances that enter crack formations and heal them without any manual human intervention. Autonomous healing is most frequently achieved by mixing concrete with fibers, crystalline admixtures, bacteria, polymers, and other such materials. Normally, chemical reactions and the presence of water make up the autonomous healing procedure that assesses natural healing mechanisms. The strength of the concrete is enhanced followed the diminished width of a given crack via the mentioned autonomous healing procedure. Contrastingly, Figure 8.1 portrays the autonomous procedures involved in artificial autonomous healing. The efficacy of self-healing is enhanced via immobilization, vascular, or encapsulation techniques that incorporate healing agents into the concrete.

Autonomously healing concrete contains crystalline admixtures. Due to the hydrophilic properties of crystalline admixtures, the water permeability reduces in the concrete while mechanical properties are enhanced, as per several academic works. Alkaline attacks serve to circumvent the corrosion of steel and the stiffness of the concrete is enhanced upon the incorporation of the crystalline admixture [5]. The permeability of the concrete diminishes due to the needle-shaped structures that arise from the crystalline admixture when it reacts with water, which in turn

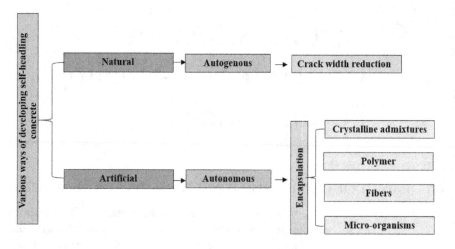

FIGURE 8.1   Flowchart of various ways to develop self-healing concrete.

functions as a waterproofing layer that also fills in the crack formations. Thus, the capacity to turn concrete into a water barrier makes its use in water tanks, dams, and other such hydraulic structures highly effective. Studies that investigate the combination of concrete with crystalline admixtures are discussed in subsequent sections.

How autonomously healing concrete is affected by crystalline admixtures is investigated in an academic work [3]. Several exposure conditions and loading-induced pre-cracks were carried out upon the concrete samples in this study. The proliferation of crack formations was halted via the use of steel fibers. The split and compressive tensile assessments also took place. Standard concrete was found to be 14% weaker in compressive strength than concrete with crystalline admixtures. The compressive strength of the concrete improved as it contained calcium sulfate-hydrate and calcite. As Table 8.1 illustrates, a 35% improvement in tensile strength was also observed.

The compatibility of crystalline admixtures as a healing agent in autonomously healing concrete was examined in another study [4]. Distinct environmental conditions were used for the mix of crystalline admixtures and the fiber-reinforced concrete. The healing performance and occurrence were measured via optic microscopic and permeability assessments. The exposure condition of immersion in water was the condition that caused the greatest decrease in crack formation width.

The self-healing efficacy of concrete mixed with crystalline admixtures was also examined in a separate paper [6]. The caliber of the autonomous healing property was evaluated via crack closing and permeability assessments. This involved an analysis of cement mixtures with 275 kg/m^3 density and a water to cement ratio (W:C) of 0.60, as well as another with 350 kg/m^3 density and a W:C ratio of 0.45, to deduce the concrete qualities. The compressive strength of standard concrete was found to be 15% weaker than precast concrete that had crystalline admixtures incorporated

TABLE 8.1

Crystalline Admixtures–Based Self-Healing Cementitious Materials

Ref.	Exposure Condition	Crack Healing	Findings
[3]	Water immersion	0.2–0.4 mm	Enhancement in compressive and tensile strength by 13.98% and 35%, respectively
[4]	Water immersion	0.86 mm	Specimens with CA shows the highest self-healing rates, for the larger crack widths up to 0.25 mm
[6]	Water immersion	0.4 mm	For cracks up to 0.40 mm, healing ratio and closing ratio were 0.99 and 0.98, after 42 days of healing, respectively

into it, as per the tests. Improved healing was demonstrated at temperatures of 30°C when the samples were immersed in water following the incorporation of crystalline admixtures.

Autonomously healing mortar surface crack formations via expansive crystalline additives in the concrete were investigated in a study [7]. The process of healing was enhanced by placing the pre-cracked sample in water for 28 days whereby said surface cracks measured around 100–400 mm in size. The filling in of the surface crack formations occurred due to the incorporation of crystalline additives (CA) and expansive additives (CSA), as per the findings. The crystalline admixture formed 1.5% of the cement weight while calcium sulfoaluminate formed 10% of the cement weight, and both combined formed the mixture. The large surface crack of 400 mm was healed in its entirety by such autonomous healing substances. Moreover, in 28 days, an absolute reduction to 0% occurred regarding the liquid penetration. Thus, a substantial enhancement in crack closing capability was triggered by the incorporation of calcium sulfoaluminate and crystalline admixtures. The calcium carbonate precipitation was crucially and positively affected by the quantity of calcium precipitated from the concrete mix. Standard concrete releases significantly less calcium ions than concrete with crystalline additives and expansive additives incorporated into it.

Concrete mixed with crystalline additives and its resultant autonomous healing performance were also investigated in other academic works [5, 8]. How the healing rate of the sample was affected by accelerated and natural curing conditions was examined. Immersion in water and air made up the natural environment conditions, while air chambers were among the accelerated environment conditions used upon the undamaged and pre-cracked beams that underwent flexural assessments. The incorporation of additives was investigated alongside samples without any additives. Pre-crack formations measuring 270 mm and 130 mm wide were created in the samples mixed with additives and the samples without. Subjected to different exposure conditions, the healing rate of the cracks 130 mm and 270 mm in the samples mixed with additives is presented in the findings.

8.2 POLYMERS AS SELF-HEALING MATERIAL

The binding of polymer chains and monomers typically lead to the formation of polymers. A polymer net layer that heals crack formations is made after polymer binds with cement particles once polymers are incorporated into the concrete. Polymers have a higher viscosity that enable them to plug in cracks efficiently. Polymers as autonomous healing agents in concrete have been studied in numerous academic works. For example, pellets filled with healing gel and made with polymers are incorporated into concrete, in a certain paper, to test the healing efficacy of such a combination [9]. The encapsulation method is used in this paper to incorporate polyurethane capsules into the crack formations as they are filled with the healing agent. The mechanical properties of the concrete sample are assessed via split tensile and three-point bending tests. Increases in spit tensile and flexural strength are observed in the samples after a week of curing. The use and amalgamation of superabsorbent polymers and polyurethane occurred in this paper [10]. Independently, the concrete also had acrylate and acrylamide superabsorbent polymers (SAPs) as well as fibers made of polyurethane mixed into it. The compressive strength of standard concrete was found to be 18% greater than concrete with superabsorbents mixed in, thereby demonstrating that improved strength is not an outcome of adding polyurethane. When subjected to a flexural test, the crack formations up to 250 mm wide in the specimens are developed. The generation of calcium carbonate is triggered by the hydration process that the swelling of superabsorbent polymers catalyzes by releasing water into the crack formations. As a result, the calcium carbonate plugs and heals the cracks.

How acrylate end-capped polymer precursors affect the healing of concrete crack formations is another area studied [11]. This takes place via the entry of capsules into the concrete containing epoxy acrylate and polyester acrylate–like chemicals. Elongation causes damage to the polymer walls after concrete specimens mixed with polyester-based substances and propylene glycol are subjected to a load. Upon being placed in an alkaline environment, such polymers still do not deteriorate, given their strong self-recovery capacity. Less than 50% is the polyester-based healing agents' strain limit. The width of crack formations rises when rising to more than 50%. The restriction of cement particles and polypropylene binding is observed via visual testing. The potential of acrylate polymers as healing agents is significant, as per the findings. Nonetheless, the type of acrylate capped precursors used determines the self-recovery procedure.

Healing adhesives contained within microcapsules that are mixed into cementitious materials is another area researched in the literature [12]. Autonomous healing ability is generated by combining cement paste with polystyrene-divinylbenzene (stn-DVB), diglycidyl ether of biphenyl, epoxy resin, and other such adhesive materials. Parameters involving strength were substantially enhanced via this mix. As illustrated in Table 8.2, the flexural strength and compressive strength of standard concrete was 1.3 times and 1.9 time weaker, respectively, than the autonomously healing concrete. The efficacy of superabsorbent polymers functioning as self-recovery agents was examined in another study [13]. The efficacy of autonomous healing was ascertained via a water flow test. Upon exposure to water, superabsorbent polymer

TABLE 8.2
Polymers-Based Self-Healing Cementitious Materials

Ref.	Method	Test	Major Findings
[9]	Encapsulation	Three-point bending and split tensile	Flexural strength was observed up to 52% and 62%, while tensile strength was increased up to 40% and 64%; polyurethane is added in tubes
[10]	Vascular	Compressive strength	A decrease in compression strength up to 18% was observed for SAPs; due to swelling of SAPs, the crack width of up to 50–250 mm was healed
[12]	Encapsulation	Compressive and flexural strength	The efficiency of strength in compression test is 1.9 times normal strength of concrete; the flexural test is 1.3 times normal strength of concrete
[15]	Encapsulation	Three-point bending	There is a 35% increase in flexural strength
[16]	Direct	Ultrasonic pulse velocity (UPV)	The results are as such based on the dosage of the polymers the healing rate is increasing; healing efficiency depends on the age of damage level in both specimens containing with or without polymers

has an expansive trait and functions as a cross-linked polymer. Large crack formations have their width significantly reduced via said self-healing procedure. Table 8.2 demonstrates that the flexural strength of standard concrete is 8% weaker than the self-healing concrete, as per the findings. Cementitious composite materials and their autonomous healing capabilities were investigated in a separate academic work [14]. The reaction rate was augmented via the accelerator microencapsulating triethyl borane and methyl methacrylate monomers. A comprehensive mixing of the solution containing a combination of microcapsules and methyl methacrylate occurred within the concrete. The preparation consisted of readying three kinds of carbon microfiber-reinforced mortar samples. The first involves only the control mortar, the second includes sulfonated polystyrene modified mortars, and the third involves the autonomously healing mortar. These three kinds of samples were analyzed to deduce their mechanical performance. As can be seen in Table 8.2, compressive strength improves as per the SHM and SPSM findings. The strength of the healed crack formations in the concrete were examined following flexible polymers being used as healing agents in an academic work [15]. The encapsulation method enabled the incorporation of polymers into the concrete. The tensile strength of the autonomously healing concrete was ascertained via split tensile tests. Following 72 hours of curing at a temperature of 20°C, the analysis took place. The flexural strength of the concrete improved by 35% after the crack formations were healed by the flexible polymers.

Concrete specimens are pre-cracked measuring 100–400 mm wide in a study whereby superabsorbent polymer (SAP) is put into the concrete [17]. At a pressure gradient of 4, the sodium chloride solution was bypassed in order to examine the crack sealing efficacy of SAPs. Standard concrete had a rate of water flow 85% higher than the samples with SAPs. Similarly, in comparison to standard concrete, a substantial reduction in the water flow was experienced by other mortar samples. Poly-acrylate was found to be inferior to SAP with regard to healing crack formations. As illustrated in Table 8.2, crack formations in concrete structures are healed more effectively when the SAP particle is larger, as per the findings. Diverse doses of acrylic polymers and styrene-butadiene rubber were mixed into autonomously healing concrete in another academic work [16]. The hydration process and healing time were evaluated via thermogravimetric and UPV assessments. Following nearly three weeks of curing, a reduction in the rate of healing time of crack formations was observed. The incorporation of styrene rubber into the concrete explicates this. Cement particles bond with the polymer, which in turn means that the polymer dosage determines the healing time. The incorporation of styrene-butadiene rubber causes a reduction in the hydration process while the healing time is augmented by the polymer content.

8.3   FIBERS AS SELF-HEALING MATERIAL

Man-made and natural fibers make up the two kinds of fibers. The former are also known as synthetic fibers, fabricated at the industrial level, and include polyvinyl alcohol, asbestos fiber, polypropylene, vinyl alcohol, polyester, and polyethylene. Drying shrinkage and plastic-induced cracks are regulated and prevented via the incorporation of synthetic fibers in concrete which also enhance the toughness and strength of the concrete. Water penetration through pores is safeguarded against as fibers decrease the permeability of the concrete mix. Enhancements to the mechanical properties of the concrete occur from the incorporation of fibers. The utilization of fiber functioning as healing agents within concrete has been explored in numerous studies. The ability of fiber-reinforced cementitious materials to seal crack formations was examined in an academic work [18]. Steel fibers, wollastonite, antifoaming agents, silica fume cement, and superplasticizer were among the materials used in this study. Under the condition of water immersion, numerous cracks in close proximity to one another were more effectively healed by said materials. Surface cracks undergo microscopic and air permeability examinations. The autonomous healing procedure commenced with immersion in water, along with the pre-cracking of the specimens. A decrease in the air permeability ratio was observed while the crack formations were filled by the precipitated crystalline particles stemming from the fiber-reinforced cementitious materials. The results indicated that the un-cracked samples demonstrated reduced air permeability in comparison with the cracked samples.

Multiple synthetic fibers were utilized to ascertain the healing performance of fiber-reinforced cementitious concrete (FRCC) in a separate paper [19]. The paper utilized several distinct fibers such as polypropylene, ethylene-vinyl alcohol, polyvinyl alcohol, and polyacetal. The primary materials consisted of polycarboxylic acid (superplasticizer), silica fume, cement, and sand. C-PP and PP, PVA-II and PVA-I, EVOH, and POM—which are 18 mm, 11 mm, 37 mm, 14 mm, 15 mm, and 48 mm in diameter, respectively—make up the properties of the fibers. Subsequently, microscopic and air permeability examinations were carried out. Precipitation near the fibers led to a reduction in water permeability, particularly in the case of PVA, and in crack formations measuring up to 0.3 mm wide; a higher rate of precipitating calcium carbonate was observed in the high polarity of the fiber composites, as per the findings. A reduction in water permeability was found in the PVA-II series samples, while healing occurred in 72 hours for the PVA series samples.

In conditions of cryogenic temperatures, fiber-reinforced concrete's ability to autonomously heal was examined [20]. For 28 days, at a temperature of 20°C, the samples were cured in water following a 72-hour period when the samples were kept at room temperature and cured under air exposure. A week was needed to seal the cracks between 10 lm and 50 lm in the pre-cracked samples within the curing period of 28 days. The sealing of cracks measuring up to 50 lm was achieved when the sample was subjected to 24 hours of cryogenic temperatures. Flexural strength is also affected by the length of the fiber, whereby the shorter the fiber, the weaker the flexural strength. Upon being subjected to 72 hours of air curing and cryogenic temperatures, an improved flexural strength was demonstrated by the pre-cracked samples—a kind of steel fiber with an aspect ratio of 97.5. The weakest energy absorption capability at crack formations was observed in the pre-cracked samples containing steel fiber with an aspect ratio of 100. To what extent mixing cellulose fiber into concrete affected water permeability and autonomous healing efficacy was examined in another paper [14]. The autonomous healing rate of the specimen was ascertained via testing the UPV while the coefficient of porousness was evaluated by performing water penetrability assessments. Upon the incorporation of 0.5% of cellulose fibers, a decrease in the compressive strength of the concrete was observed. Contrastingly, the flexural strength of standard concrete was found to be 7.84% weaker than the sample, while the latter also displayed better tensile strength.

The incorporation of microfibers and SAPs into the concrete mix and how this influenced the mechanisms of autonomous healing was examined in another study [21]. The capability to seal cracks was improved following the incorporation of SAPs. Upon exposure to a wet environment, SAPs can swell, which explicates the improvement. When subjected to wet and dry cycles, a crack formation 138 mm in size was healed upon 1% of SAP being included in the specimens. The incorporation of SAP into concrete led to the healing of cracks and a decrease in water permeability. Enhancing the efficacy of autonomous healing via the incorporation of diverse bacteria and fibers into the concrete was examined in a separate academic work [22]. A decrease in the concentration and growth of the bacteria was observed while crack formations measuring 500 lm in width were successfully sealed. Following one week of curing, the cracked control samples containing no bacteria or fiber demonstrated 30% healing, and after 28 days of curing, the figure improved by 2%. Following 28 days of curing, the control samples demonstrated a higher crack repair ratio than samples manufactured with PP1 (polypropylene) fiber. After 28 days of curing, a crack-healing ratio of 31% was observed in sample containing the PP2 fiber, while the first week of curing demonstrated a 36% improvement in the ratio. The sole incorporation of PP1 and PP2 fibers into the mortar demonstrated inferior findings to the incorporation of the polyvinyl alcohol (PVA) fiber into the mortar. A crack-healing ratio of 44% rose to 50% in the mortar containing PVA fibers following seven days and 28 days of curing, respectively. Enhanced healing rates were observed when microscopic organisms, otherwise named bacteria, were mixed in the mortar. Samples solely containing fiber demonstrate lower healing rates than samples with fiber and bacteria. Samples with such a mixture of fiber and bacteria in the concrete demonstrated diminished absorption of water and improved flexural strength.

8.4   SELF-HEALING PERFORMANCE

How the efficacy of autonomously healing concrete is affected by fibers, polymers, and diverse crystalline materials is examined extensively in the literature. The mechanical properties of autonomously healing concrete were ascertained via multiple assessments such as the split tensile test, the water tightness test, the flexural test, the permeability test, and the compressive test. Identifying and discovering the calcium formation with concrete crack formations mixed with artificial materials was carried out via EDX, SEM, and similar comprehensive analyses. Diminished water permeability was achieved via the surface layer cracks in the concrete being filled in after reactions were hastened through crystalline admixture use. Despite experiencing air exposure, such samples can seal their crack formations as the moisture in the environment can be absorbed as the admixtures have hydrophilic properties. An enhanced crack sealing capacity develops when crack formations are exposed to water [3]. Upon water exposure, such hydrophilic products melt and recrystallize within the crack formations, after the incorporation of crystalline admixtures. The internal crack formations of the sample are plugged in after calcium carbonate is produced as the disintegration of ca^{2+} is facilitated by the incorporation of expansive additives and crystalline admixtures into the concrete. Exposure to humid content in the environment enhances the precipitation of calcium, as the additives mixed into the concrete have a higher pH [7]. The incorporation of multiple polymer precursors into the concrete enables larger and wider crack formations of the sample to be sealed. The mechanical properties of the concrete were enhanced due to the superior binding of mortar and the polymers that was facilitated by the elastic properties of polyurethane [15].

The microcapsules that functioned as healing agents contained triethylborane and methyl methacrylate monomer. After 28 days of curing, a 45.8% rise in compressive strength was observed following the incorporation of the microcapsules in the concrete [20]. As a result of swelling, the polymer particles become larger, which leads to a reduction in the micro-cracks within the concrete following the use of SAPs in the concrete. Nonetheless, high pH environments cause a reduced swelling of SAPs [13]. Autonomous healing of the concrete was also carried out by utilizing fibers. However, a reduction in the compressive strength of the concrete occurred due to the use of fibers such as cellulose [23]. The mix ratio, the cracked surface, the roughness, and the reaction rate of fibers with concrete are among the influential aspects that must be considered [19]. Autonomous healing capability is enhanced via the utilization of ultra-high performance hybrid fiber-reinforced (UHFR) cementitious concrete [18]. The incorporation of bacteria and polypropylene into concrete was found to be inferior in terms of healing efficacy to PVA-based concrete [22]. The sealing of crack width was enhanced through the incorporation of microfibers and SPAs into the concrete [21]. The autonomous healing capability and flexural performance were also enhanced upon the incorporation of steel fibers into the concrete while being subjected cryogenic temperatures [20].

8.5 SUMMARY

The conclusions of this chapter are as follows:

i. The results indicated that using polymers and crystalline admixtures gives better performance in the healing of cracks and increased in strength of concrete.
ii. Concrete cracks width up 0.93 mm are healed by adding crystalline admixtures and curing underwater immersion.
iii. The combination of fiber-based and bacteria-based self-healing led to enhancement of the flexural strength performance.

REFERENCES

1. Asaad, M.A., et al., *Enhanced corrosion resistance of reinforced concrete: Role of emerging eco-friendly Elaeis guineensis/silver nanoparticles inhibitor.* Construction and Building Materials, 2018.**188**: p. 555–568.
2. Asaad, M.A., et al., *Improved corrosion resistance of mild steel against acid activation: Impact of novel Elaeis guineensis and silver nanoparticles.* Journal of Industrial and Engineering Chemistry, 2018.**63**: p. 139–148.
3. Reddy, T.C.S., and A. Ravitheja, *Macro mechanical properties of self healing concrete with crystalline admixture under different environments.* Ain Shams Engineering Journal, 2019.**10**(1): p. 23–32.
4. Roig-Flores, M., et al., *Self-healing capability of concrete with crystalline admixtures in different environments.* Construction and Building Materials, 2015.**86**: p. 1–11.
5. Reddy, C.M.K., B. Ramesh, and D. Macrin, *Effect of crystalline admixtures, polymers and fibers on self healing concrete-a review.* Materials Today: Proceedings, 2020.**33**: p. 763–770.
6. Roig-Flores, M., et al., *Effect of crystalline admixtures on the self-healing capability of early-age concrete studied by means of permeability and crack closing tests.* Construction and Building Materials, 2016.**114**: p. 447–457.
7. Sisomphon, K., O. Copuroglu, and E. Koenders, *Self-healing of surface cracks in mortars with expansive additive and crystalline additive.* Cement and Concrete Composites, 2012.**34**(4): p. 566–574.
8. Krelani, V., and L. Ferrara, *Self-healing capacity of concrete with crystalline additives: Natural vs. accelerated exposure conditions.* Proceedings of the 4th international conference on self-healing materials (ICSHM). 2013: Ghent, Belgium. p. 426–430.
9. Van Tittelboom, K., et al., *Self-healing efficiency of cementitious materials containing tubular capsules filled with healing agent.* Cement and Concrete Composites, 2011.**33**(4): p. 497–505.
10. Van Tittelboom, K., et al., *Comparison of different approaches for self-healing concrete in a large-scale lab test.* Construction and Building Materials, 2016.**107**: p. 125–137.
11. Araújo, M., et al., *Acrylate-endcapped polymer precursors: Effect of chemical composition on the healing efficiency of active concrete cracks.* Smart Materials and Structures, 2017.**26**(5): p. 055031.

12. Li, W., et al., *Self-healing efficiency of cementitious materials containing microcapsules filled with healing adhesive: Mechanical restoration and healing process monitored by water absorption.* PLoS One, 2013.**8**(11): p. e81616.

13. Gruyaert, E., et al., *Self-healing mortar with pH-sensitive superabsorbent polymers: Testing of the sealing efficiency by water flow tests.* Smart Materials and Structures, 2016.**25**(8): p. 084007.

14. Yang, Z., et al., *Laboratory assessment of a self-healing cementitious composite.* Transportation Research Record, 2010.**2142**(1): p. 9–17.

15. Feiteira, J., E. Gruyaert, and N. De Belie, *Self-healing of moving cracks in concrete by means of encapsulated polymer precursors.* Construction and Building Materials, 2016.**102**: p. 671–678.

16. Abd_Elmoaty, A.E.M., *Self-healing of polymer modified concrete.* Alexandria Engineering Journal, 2011.**50**(2): p. 171–178.

17. Lee, H., H. Wong, and N. Buenfeld, *Self-sealing of cracks in concrete using superabsorbent polymers.* Cement and Concrete Research, 2016.**79**: p. 194–208.

18. Kwon, S., et al., *Experimental study on self-healing capability of cracked ultra-high-performance hybrid-fiber-reinforced cementitious composites.* 3rd International conference on sustainable construction materials and technologies, Kyoto, Japan, 2013.

19. Nishiwaki, T., et al., *Experimental study on self-healing capability of FRCC using different types of synthetic fibers.* Journal of Advanced Concrete Technology, 2012.**10**(6): p. 195–206.

20. Kim, S., et al., *Self-healing capability of ultra-high-performance fiber-reinforced concrete after exposure to cryogenic temperature.* Cement and Concrete Composites, 2019.**104**: p. 103335.

21. Snoeck, D., et al., *Self-healing cementitious materials by the combination of microfibers and superabsorbent polymers.* Journal of Intelligent Material Systems and Structures, 2014.**25**(1): p. 13–24.

22. Feng, J., Y. Su, and C. Qian, *Coupled effect of PP fiber, PVA fiber and bacteria on self-healing efficiency of early-age cracks in concrete.* Construction and Building Materials, 2019.**228**: p. 116810.

23. Singh, H., and R. Gupta, *Influence of cellulose fiber addition on self-healing and water permeability of concrete.* Case Studies in Construction Materials, 2020.**12**: p. e00324.

9 Encapsulation Technology-Based Self-Healing Cementitious Materials

9.1 INTRODUCTION

The impact of diverse load and non-load elements, as well as the somewhat lower tensile strength of concrete, makes the formation of cracks unavoidable. Rebar corrosion, drying shrinkage, external loading, plastic shrinkage, thermal stress, or a combination of the mentioned factors lead to the formation of cracks. This is exemplified by shrinkage causing micro-crack formations—yet upon the presence of an external load, cracks proliferate into a network at a lower stress level. The chemical degradation of concrete can then occur as chemicals and moisture make ingress into the concrete through said network of cracks. High levels of precipitation and moisture in the tropics aggravate such issues. Substantial costs, complicated accessibility, and the production of pollution are among the issues linked with manual interventions to repair cracks. Currently, a range of cement-based and chemical repair materials are utilized. Environmental and health risks, as well as material unsuitability, are issues linked with using chemical healing agents, while 7% of the anthropogenic CO_2 emissions worldwide are linked to cement production [1–3]. Significant structural or durability problems frequently arise as micro-cracks go unperceived when they form at the initial or later stages of construction using concrete. Recurrent maintenance on concrete structures could prevent this, yet it exacts a significant financial toll. Thus, making manual intervention obsolete via a sustainable cost-efficient method of repairing crack formations is necessary.

The use of autonomous healing techniques in building structures has drawn much acclaim over the last decade due to their capacity to repair deterioration within materials of a high caliber. Thus, additional economic and environmental benefits can occur as self-healing techniques are implemented that relieve the need for maintenance operations. Academic works have investigated the precipitation of calcium carbonate provoked by a microbial as a sustainable manner of plugging in crack formations. *Bacillus subtilis* is one such bacteria and autonomously healing microbe that directly acts upon a calcium compound like calcium lactate to fill in crack formations via the precipitation of calcium carbonate [1]. Such precipitation also occurs through a ureolytic bacteria like *Bacillus sphaericus*, causing the decomposition of urea [4–5]. The setting of concrete occurs in an environmentally sustainable way during the process of microbe-induced calcium carbonate precipitation in the concrete [6]. For example, there is no health risk associated with *Bacillus sphaericus* [7].

DOI: 10.1201/9781003195764-9

Moreover, the likelihood of reinforcement corrosion is decreased as oxygen is consumed in the mentioned procedure. The bacteria species type *Bacillus* is revealed to be an optimal autonomous healing agent as it has the ability to form spores while also being resistant to highly moist and alkaline environments. Thus, the bio-agent for calcite precipitation within the literature [8–9] is most frequently *Bacillus*.

9.2   BIO-BASED HEALING AGENTS: APPROACHES AND MECHANISMS

In optimal circumstances, the availability of nucleation sites, solution pH, calcium ions concentration, and the dissolved inorganic carbon concentration affect the precipitation of calcium carbonate in a natural environment [10]. The first of these aspects is delivered by the bacterial cell, while the other three are linked to the concrete matrix. Undergoing hydrolysis of urea by bacterial metabolism or the conversion of calcium compounds like calcium lactate are ways in which to fulfill bacterial precipitation. The first method involves the hydrolysis of urea into carbonate and ammonium, which triggers calcium carbonate precipitate. The enzyme urease functions as a catalyst during hydrolysis, and it is generated by the *Bacillus sphaericus* bacteria. The generated carbonate reacts with calcium ions extracted from calcium nitrate or other calcium sources by the negatively charged bacterial cell, resulting in calcium carbonate precipitation. The second method involves the production of carbon dioxide and calcium carbonate after calcium lactate reacts with the oxygen that enters into the concrete via crack formations on the surface [10]. Additional calcium carbonate is generated, which can be used for autonomous healing, via the reaction of carbon dioxide with any portlandite particles in close proximity. Non-hydrated calcium hydroxide particles are still present when concrete is fresh; thus, this method is more effective in such a context.

Hence, the long-term lifespan of bacteria, the crack depth or width to be repaired, moisture, and the age of concrete are among the elements that affect the efficacy of autonomous healing in concrete, yet such healing can be achieved via any of the mentioned methods. Improving autonomously healing concrete was attempted via the introduction of bacteria spores in different contexts to deduce its compatibility [11]. *Bacillus pseudofirmus*, *Bacillus halodurans*, and *Bacillus cohnii* were the three kinds of bacteria that were assessed and mixed in the cement stone. The tensile and compressive strength of cement stone chips were assessed after they were cured in a peptone-based medium and yeast extract. Variance between the control samples and samples including bacteria were insignificant. Following 12 days of incubation, precipitation of calcium carbonate crystals were illustrated via scanning electron microscopy (SEM) images. Nonetheless, germination occurred after specimens were cured in the medium, once given food in the form of organic carbon sources which were externally sourced.

More recently, an academic work investigated the potential impacts of mineral precursor compounds, peptone, and yeast extract [8]. The highly alkaline environment of concrete is the principal difficulty that the bacteria must overcome to survive, and thereby commence autonomous healing. The promise of direct additions to the concrete of different kinds of bacteria-induced carbonate precipitation that demonstrated suitability throughout was studied in other academic works [12]. *Bacillus cohnii* and *Bacillus pseudofirmus* were assessed and are species of spore that form alkaliphilic bacteria. The mineral precursor compound was determined by assessing the calcium lactate and calcium acetate. A marginal improvement to strength occurred via the calcium lactate, while the calcium acetate affected the strength. The 28-day specimens did not have precipitate on them, while the 7-day samples had calcium carbonate crystals between 20–80 mm in size. Decreased pore size, along with concrete's highly alkaline environment diminishing the number of suitable bacteria spores, could explain this, as per another study [13]. Nonetheless, the deterioration of the concrete matrix due to nitric acid, generated by increased metabolism activity, or reinforcement corrosion due to excess ammonia production which occurs in the urease-based method demonstrates why the aforementioned technique is considered superior by some.

The direct incorporation to concreate of the substrate and alkali-resistant spores that form bacteria were investigated with regard to the influence of crack formation age, curing condition, and crack formation width. Wet-dry cycles at 25°C, water curing, and wet curing were among the post-cracking incubation conditions attempted following the presentation of crack formations between 0.1 mm and 0.5 mm wide. Crack formation ages of 90, 60, 28, 14, and 7 days were used to assess the area repair rate of such diverse cracks and thereby ascertain the healing performance. Roughly 85% of crack formations 0.1–0.3 mm wide were filled in entirely, while crack formations up to 0.3 mm wide were repaired entirely when subjected to the water curing condition. While the wet-dry cycle condition demonstrated a slower rate of healing, said condition, along with water curing were found to produce the best healing results. Older crack formations were not healed as effectively as new or young crack formations. The shorter transport distance of the mineral that stems from a reduction in concrete porosity partially explains this, while the lack of a protective shell for the bacteria, which results in a low survival rate, is also relevant [13].

The binding of the original matrix with mineral precipitate is a key area to investigate, along with deducing the latter's properties to heighten autonomous healing efficacy. The binding of precipitated calcium carbonate with the parent concrete, as well the former's nanomechanical properties, were ascertained via a non-indentation technique by certain academics, following the addition of a non-ureolytic bacteria within the mixing water that revealed the mechanical performance of autonomously healing concrete [14]. The creation of crack formations led to the addition of air pockets for bacteria survival and a decrease in the alkalinity of the cement paste, while the addition of air entraining agents, silica fume, and basalt fibers sustained the integrity of the specimens, thus demonstrating the ground-breaking nature of this technique. Given that durability and mechanical properties of concrete can be affected by air voids in the tropics, this method may not be as suitable as in temperate climates where the dangers of thawing and freezing are reduced by such air voids. Nonetheless, in the case of a fire, the release of vapor pressure is enabled by these voids, thereby enhancing the concrete's resistance to fire. The role of the mineral precursor compound was experimentally filled by calcium glutamate and calcium lactate. Calcium lactate produced a less substantial conversion of calcium ion to carbonate than calcium glutamate, where the latter formed granular-shaped crystals which could not be seen in the former. Nevertheless, while thicker and denser carbonate from glutamate was found in the transition zone, the binding of carbonates made from lactate and glutamate was found to vary insignificantly. Calcium glutamate specimens were observed to demonstrate a greater capacity in restoring flexural strength. Bacterial metabolism triggered mineral precipitates that filled in part of the crack formations between 0.1 mm and 0.4 mm wide. The efficacy of external healing was not surpassed by samples with bacterial spores, yet the latter did exceed the healing performance of control specimens. The limited amount of nutrients explicates this, thereby affecting the performance of the bacteria despite the voids aiding the bacteria.

The production costs of cultures such as the axenic spores of bacteria are substantial, which is the principal issue of the approach that demonstrates promising results regarding effective self-healing. The microbial community named Cyclic EnRiched Ureolytic Powder (CERUP) aimed to address this issue by offering a cheaper bio-agent alternative [15]. Reduced production costs were achieved by making the specimen in non-sterile conditions while using a sub-stream of a vegetable plant as the microbial source. During a period of four weeks, the capacity to plug in crack formations and autonomously heal was assessed via the incorporation of non-autoclaved and autoclaved CERUP. The largest crack formation repaired by the autoclaved CERUP was just under 0.37 mm wide, while the largest crack formation healed by the non-autoclaved CERUP was 0.45 mm wide. A more accurate representation of the autonomous healing capacity of the non-sterile agent should be deduced via additional evaluations regarding durability and strength to ascertain its proficiencies—besides significantly lower costs—and its deficiencies.

9.3   INCORPORATION OF ENCAPSULATED HEALING AGENTS

Autonomous healing occurring during the entire lifespan of a structure is indicative of a highly effective self-healing technique that can fill in crack formations indefinitely. Thus, a critical aspect is survivability of the bacteria. Nonetheless, multiple obstacles hindering survivability of bacteria can appear upon the direct addition of bio-agents to the concrete. Relatively young specimens were found to be the only kind that enacted efficient autonomous healing as the unprotected spores had a lifespan of two months, as per an academic work [16]. The hydration of cement, the mixing of concrete, and the alkalinity of the cement matrix are among the elements that cause contribute to such a short lifespan. Residing in a highly alkaline environment for a prolonged period of time could significantly reduce the activity of spores. Additionally, the impact of aggregates along with the mixing force can lead to spores being damaged in the mixing stage. Despite the average size of bacteria cells being above 0.5 mm, a reduction in the pore size of the matrix and porosity is triggered by the hydration of cement up to 0.5 mm [4].

Thus, at an advanced stage, a concrete structure can experience a significant or even absolute decrease in the germination of cells due to the shrinkage of pores. The bacteria-induced carbonate precipitation and the concrete properties are not affected when the bacteria is encapsulated, thereby working around the mentioned limitation. Hydrogel encapsulation, diatomaceous earth, melamine-based microcapsules, polyurethane and silica gel in glass tubes, and expanded clay aggregate encapsulation techniques to safeguard the bacteria have been used in the literature [17–18].

Bacillus sphaericus spores were encapsulated via melamine-based microcapsules in an academic work [4]. Urea and yeast extract were among the nutrients mixed in the concrete, along with the mineral precursor calcium nitrate. The ratio between the initial cracked area and the healed crack area was used to evaluate the healing efficacy of the bacteria. The healing ratio could not surpass 50% in the specimens with no spores, while those with encapsulated spores achieved a ratio between 48% and 80%. When water functioned as a medium during the wet and dry cycles that the samples underwent, the greatest decrease in crack area occurred. One drawback is that manual intervention may be required under typical circumstances when performing the wet cycle, as the procedure takes around 16 hours. A crack formation close to 1 m wide was the greatest width repaired. While a capsule dosage of 5% led to less deviation, the most effective dosage for crack healing and restoration of water permeability was 3%. The 5% dosage was not considered superior, as a greater decrease in concrete strength was observed for said dosage than the former.

Little to no manual intervention is needed, and a source of internal moisture for activity and growth of the bacteria is delivered in the inventive hydrogel encapsulation technique, undertaken by certain academics [10]. The bacteria is also safeguarded in the protective shell of the capsule. When the hydrogel contained both bioreagents and the encapsulated bacteria, healing efficacy reached between 40% and 90%. Additionally, the highest reduction of 68% water permeability occurred. The retention ability of hydrogels and water uptake enabled the hydrogel encapsulated bacteria to perform healing effectively. At a temperature of 20°C in an air composition with 60% relative humdity, pure hydrogel after a day and half a day retained 30% and 70% of water absorbed, respectively [10]. Bacterial activity can be encouraged in tropical climates by the effective retention and absorption of water that hydrogels offer, given that such climes experience high levels of precipitation and humidity. Nonetheless, whether the hydrogel is non-ionic or ionic is a relevant factor [10]. While retention and water uptake ability is not affected by the ions in moisture via non-ionic hydrogel, ionic hydrogels and their uptake ability can be influenced by the chemicals in the air due to being pH responsive. The curing condition of dry and wet cycles lasting eight hours and 16 hours, respectively, was found to be the most effective regarding healing performance upon the use of melamine-based microcapsules [4]. A substantial decrease in contact time with water can occur when utilizing hydrogel. Moreover, the use in construction of concrete containing hydrogel offers the benefits of replacing the need to manually add water, given that moisture from the air can be absorbed by the hydrogel.

Whether modified sodium alginate-based hydrogel suitably functioned as a bacterial spore carrier was assessed in another academic work [18]. The consumption of oxygen at a damaged site in the concrete sample served to evaluate suitable bacteria spores following hydrogel encapsulation. While the best majority of spores maintained their encapsulated form, the concrete mixing stage led to a leakage of around 1% of spores of the altered hydrogel. Upon the incorporation of 0.5% and 1% of the hydrogel mass, reductions of 30% and 23.4% were experienced in the compressive and tensile strength, respectively, despite the workability of the concrete not being notably affected. The incorporation of hydrogel could have led to macro voids forming, which could explicate the negative impacts mentioned. Diminished strength and the production of voids in concrete is an outcome of SAP, while contrastingly, it also aids development of strength and provides moisture for internal healing. The age of the concrete, the dosage of SAP, and the ratio of water to cement (W:C) determine which of the contrasting effects outweigh the others [19]. A decrease in strength and less strength development, upon incorporation of SAP, are likely when the W:C is more than 0.45 in the context of the gel-to-space ratio premise. Specimens with SAP may have experienced diminished strength, as the academic work mentioned utilized a W:C of 0.5, which is a high value[18].

The precursor compound for mineral precipitate, calcium lactate, was used alongside porous expanded clay aggregate which immobilized bacteria spores, as per an older study [19]. Exposure to air incites bacterial action, which in turn induces calcium carbonate precipitation following the fracturing of soft and lightweight clay aggregates. A crack formation 0.46 mm wide was the maximum width of a filled in crack following the specimens' immersion in tap water for 14 days. No signs of decreased viability of the bacteria were observed after six months. Bacterial precipitation activity is encouraged by the internal source of moisture provided by clay aggregates, which are also prevalently used in lightweight concrete. Nonetheless, pore structure, expanded clay aggregate spacing, and water in aggregate are factors that affect the efficacy of clay aggregates [1]. The decrease in strength is the principal drawback in using clay aggregates rather than the standard granite aggregates. The compressive strength of concrete depends upon the toughness of the aggregate, given the vast majority of concrete is made up of the aggregate in standard concrete. The aggregate is stronger than the matrix, where cracks appear eventually, as strong aggregates resist crack formations in standard concrete. Nevertheless, the aggregate may be less tough than the matrix, in the case of clay aggregates, making it highly probable that crack formations breaking the aggregate. Thus, crack formations are attracted to the weaker plane that soft aggregates generate. Unsuitability for structural use can be inferred, as up to a 50% decrease in strength was observed after 28 days when incorporating lightweight clay aggregates [17].

9.4   EVALUATION OF BIO-BASED SELF-HEALING SYSTEMS

Currently, adding bio-agents exacts a high cost. However, one way to make such agents commercially feasible is to increase the lifespan and resistance of autonomous healing agents to effectively perform in harsh environments and endure the multiple cycles of loading that structures frequently experience. Another strategy is to reduce the production costs of bio-agents. During the lifespan of a structure, persistent deterioration makes multiple maintenance operations necessary, which in turn exacts high costs. A life cycle cost model formulated by van Breugel explicates this issue [20]. The cost incurred to build a structure may even be surpassed by the cost to repair the structure. Contrastingly, while the initial construction cost may increase by using autonomously healing concrete, the total cost of the structure throughout its lifespan will decrease drastically as damages are dealt with autonomously to no extra cost.

One of the critical components of autonomous healing is capsule design. Identifying and producing capsules that hardly—if at all—affect concrete properties should be an area of further study. Additionally, when under fatigue, there are scant findings on bio-based autonomous healing performance. Capsule release behavior determines the bio-based autonomously healing concrete's fatigue performance. Smart or controlled release from capsules can enable autonomous healing to occur when subjected to multiple loading cycles. Reduced alkalinity of concrete matrix leads to corrosion of reinforcement bars, and an academic work aimed to heal such deterioration via the smart release of microcapsules [21]. Nonetheless, implementing the smart release of capsules for bio-based autonomous healing is a method in need of further replication and further study to be validated. The incorporation of capsules provokes the formation of weak spots in the mortar, yet such spots can be diminished in size by nano-capsules. While only the encapsulation of a chemical healing agent has been carried out in the present, the sonification method enabled the creation of urea-formaldehyde capsules that measure 77 nm thick and 220 nm in diameter [22]. Biological healing agents could be integrated into this approach if the capsule material can be altered to suit bio-agents. Nevertheless, the formations of cracks in the matrix can occur in nanoparticle agglomeration spots; thus, non-particle debris must be assessed to guarantee the absence of agglomeration.

A lengthy period of between 14 and 21 days is needed for biological agents in the concrete with appropriate curing conditions to enact self-healing of crack formations, as per the results. Bacteria in the alkaline concrete environment produce a lower precipitation rate, explaining the ample time taken. Higher precipitation rates and increased bacteria lifespans could be achieved via the creation of genetically modified bacteria cultures following further interdisciplinary investigations. In such a case when further investigations produce results, quicker healing of wider crack formations may be possible. Through biological action, quicker and more effective autonomous healing can be attained by placing an emphasis on the vital aspect of controlling the width of crack formations [23]. Following autonomous healing, highly effective restoration of original properties can be achieved via incorporating hybrid fibers to regulate width of cracks [24–25]. Healing agents and products can place themselves close to the faces of crack formations via the amalgamation of polymer fibers and steel that also act to hinder the width of cracks [24].

Conducting further investigations in a real-life construction site is required to assess autonomous healing performance more accurately. Analyzing the social and environmental benefits, diminishing costs, and improving the lifespan of concrete structures should be among the research objectives of future studies. How climate change acclimatization is affected by bio-based autonomously healing concrete is another relevant area for investigation [26]. The life cycle of bio-based autonomously healing systems can be enhanced via the implementation of life cycle testing methods and other sustainability evaluation techniques [27]. Autonomously healing composites that are more sustainable can be generated by mixing carbon sequestering materials with biological healing agents [28]. Finally, bio-based autonomous healing performance in actual buildings can be evaluated more accurately via the establishment of assessment standards and other stipulations. The application of bio-based autonomously healing concrete and other such materials is a distinct possibility in the near future, as much academic attention has been placed in the field over the last decade. Nonetheless, prevalent industrial use can only occur once the present technological challenges are overcome.

9.5 SUMMARY

The following conclusions can be drawn from this chapter:

i. It is a fact that the initial cost of incorporating bio-agents may be high at present. There may be two possible ways to make it commercially viable: by reducing the cost of production of bio-agents, or by designing the self-healing action for longer life so that it works well under multiple cycles of loading and adverse environment conditions to which structures are often subjected.

ii. Design of capsules is an integral part of self-healing.

iii. Reported results indicated that crack closure time for bio-based healing with suitable curing conditions takes a long time, which typically spans at least 14–21 days.

REFERENCES

1. Gupta, S., S. Dai Pang, and H.W. Kua, *Autonomous healing in concrete by bio-based healing agents—A review*. Construction and Building Materials, 2017.**146**: p. 419–428.
2. Worrell, E., et al., *Carbon dioxide emissions from the global cement industry*. Annual Review of Energy and the Environment, 2001.**26**(1): p. 303–329.
3. De Muynck, W., N. De Belie, and W. Verstraete, *Microbial carbonate precipitation in construction materials: a review*. Ecological Engineering, 2010.**36**(2): p. 118–136.
4. Wang, J., et al., *Self-healing concrete by use of microencapsulated bacterial spores*. Cement and Concrete Research, 2014.**56**: p. 139–152.
5. Wang, J., et al., *Use of silica gel or polyurethane immobilized bacteria for self-healing concrete*. Construction and Building Materials, 2012.**26**(1): p. 532–540.
6. Pei, R., et al., *Use of bacterial cell walls to improve the mechanical performance of concrete*. Cement and Concrete Composites, 2013.**39**: p. 122–130.
7. Luna-Finkler, C.L., and L. Finkler, *Bacillus sphaericus and Bacillus thuringiensis to insect control: Process development of small scale production to pilot-plant-fermenters*. 2012: INTECH Open Access Publisher, Brasil. p. 613–626.
8. Wang, J.-Y., N. De Belie, and W. Verstraete, *Diatomaceous earth as a protective vehicle for bacteria applied for self-healing concrete*. Journal of Industrial Microbiology and Biotechnology, 2012.**39**(4): p. 567–577.
9. Wang, J., et al., *Application of hydrogel encapsulated carbonate precipitating bacteria for approaching a realistic self-healing in concrete*. Construction and Building Materials, 2014.**68**: p. 110–119.
10. Hammes, F., and W. Verstraete, *Key roles of pH and calcium metabolism in microbial carbonate precipitation*. Reviews in Environmental Science and Biotechnology, 2002.**1**(1): p. 3–7.

11. Jonkers, H.M., and E. Schlangen, *Self-healing of cracked concrete: a bacterial approach.* Proceedings of FRACOS6: Fracture mechanics of concrete and concrete structures. Catania, Italy, 2007. p. 1821–1826.
12. Jonkers, H.M., et al., *Application of bacteria as self-healing agent for the development of sustainable concrete.* Ecological Engineering, 2010.**36**(2): p. 230–235.
13. Luo, M., C.-X. Qian, and R.-Y. Li, *Factors affecting crack repairing capacity of bacteria-based self-healing concrete.* Construction and Building Materials, 2015.**87**: p. 1–7.
14. Xu, J., and W. Yao, *Multiscale mechanical quantification of self-healing concrete incorporating non-ureolytic bacteria-based healing agent.* Cement and Concrete Research, 2014.**64**: p. 1–10.
15. Da Silva, F.B., et al., *Production of non-axenic ureolytic spores for self-healing concrete applications.* Construction and Building Materials, 2015.**93**: p. 1034–1041.
16. Jonkers, H.M., *Bacteria-based self-healing concrete.* 2021: In-Genium, Delft, Netherlands. p. 1–5.
17. Wiktor, V., and H.M. Jonkers, *Quantification of crack-healing in novel bacteria-based self-healing concrete.* Cement and Concrete Composites, 2011.**33**(7): p. 763–770.
18. Wang, J., et al., *Application of modified-alginate encapsulated carbonate producing bacteria in concrete: A promising strategy for crack self-healing.* Frontiers in Microbiology, 2015.**6**: p. 1088.
19. Jensen, O.M., *Use of superabsorbent polymers in concrete.* Concrete international, 2013.**35**(1): p. 48–52.
20. Van Breugel, K., *Is there a market for self-healing cement-based materials.* Proceedings of the first international conference on self-healing materials, 2007.
21. Dong, B., et al., *Smart releasing behavior of a chemical self-healing microcapsule in the stimulated concrete pore solution.* Cement and Concrete Composites, 2015.**56**: p. 46–50.
22. Blaiszik, B., et al., *Microcapsules filled with reactive solutions for self-healing materials.* Polymer, 2009.**50**(4): p. 990–997.
23. Souradeep, G., and H.W. Kua, *Encapsulation technology and techniques in self-healing concrete.* Journal of Materials in Civil Engineering, 2016.**28**(12): p. 04016165.
24. Homma, D., H. Mihashi, and T. Nishiwaki, *Self-healing capability of fibre reinforced cementitious composites.* Journal of Advanced Concrete Technology, 2009.**7**(2): p. 217–228.
25. Gupta, S., *Development of high strength self compacting mortar with hybrid blend of polypropylene and steel fibers.* International Journal of Engineering and Technology, 2014.**4**(10): p. 571–576.
26. Kua, H.W., and S. Koh, Sustainability science integrated policies promoting interaction-based building design concept as a climate change adaptation strategy for Singapore and beyond. In *Green growth: Managing the transition to a sustainable economy.* 2012: Springer, London. p. 65–85.
27. Kua, H.W., and S. Kamath, *An attributional and consequential life cycle assessment of substituting concrete with bricks.* Journal of Cleaner Production, 2014.**81**: p. 190–200.
28. Gupta, S., and H.W. Kua, *Factors determining the potential of biochar as a carbon capturing and sequestering construction material: Critical review.* Journal of Materials in Civil Engineering, 2017.**29**(9): p. 04017086.

10 Applications, Future Directions, and Opportunities of Self-Healing Concrete

10.1 INTRODUCTION

Currently, the entire planet is at risk due to the continual climate change [1–3]. The recorded increase in average temperature across the world in the past 100 years, and associated changes attributed to this, are known as global warming. Many scientists are convinced by the published evidence that this change is anthropogenic and resulted from the elevated emission levels of the global greenhouse gases (GHGs) [4–5]. Gases such as water vapor, carbon dioxide, methane, nitrous oxide, and ozone are responsible for the absorption and emission of thermal radiation. These changes in the relative quantities of the GHGs induce a proportional change in the amount of preserved solar energy. Presently, the accepted indicator for global warming is the sustained rise in the mean temperature worldwide. This definition is designed to account for the fact that there may be some localized exceptions to this rise. For example, there may be cooling experienced in a region while the global temperature may increase altogether, hence the need for use of average temperature. A key concern with the GHGs trapping of more heat in the atmosphere is that it affects both climate and short-term weather patterns. Consequently, it results in greater numbers of adverse weather events such as storms, heat waves, cold snaps, droughts, and fires [6]. Climate-related risks to health, livelihoods, food security, water supply, human safety, and economic growth are projected to increase with global warming of 1.5°C [7] and further increase of 2°C. In addition, the risks to global aggregated economic growth due to the climate change impacts are projected to be lower at 1.5°C than at 2°C by the end of this century.

For ages, ordinary Portland cement (OPC) has served as the primary structural material in the construction sector and widely used as concrete binder worldwide [8–9]. Although it is well known that large-scale manufacturing of OPC causes serious environmental pollution in terms of considerable amounts of GHGs emission, no substitute for OPC has emerged [10]. OPC production alone is accountable for nearly 6–7% of total CO_2 emissions [11]. In recent times, self-healing cementitious materials have been introduced as a new construction material to replace traditional concrete in the construction industry [3, 12]. Several notable merits of self-healing cementitious materials—such as enhancement of strength properties, reduction of

DOI: 10.1201/9781003195764-10   **133** environmental pollution, enhanced durability, energy saving attributes, and long lifespan—make them advantageous over other construction materials [3].

Repair or rehabilitation is the major concern regarding several deteriorated concrete structures around us [13–14]. Such repairs for concrete structures are necessary to assure their service lifetimes. Over the years, several repair materials are developed for concrete structures, including cement-based materials, polymers, latex, etc. [5–6]. Lately, self-healing cementitious materials revealed tremendous prospects toward emergency repairs and coating [12]. Several studies [15–18] have shown that cracked concrete has the ability to heal itself over time when exposed to water. It has been found that there is a gradual reduction in permeability of damaged concrete as water is allowed to flow through the cracks. This decrease in permeability is due to diminishing crack widths as healing occurs. A number of approaches to induce self-healing in concrete have been attempted. Of these approaches, the most common are chemical encapsulation [19], bacterial encapsulation [20], mineral admixtures [21], chemicals in glass tubing [22], and intrinsic healing with self-controlled tight crack widths [23].

10.2 SUSTAINABILITY OF SMART MATERIALS-BASED SELF-HEALING CONCRETE AND NANOTECHNOLOGY

Cementitious self-healing materials and nanotechnology are two of the most significant scientific and industrial breakthroughs of the twenty-first century, offering great advantages toward concrete sustainability in construction fields such as energy storage, high performance, corrosion resistance, environmental remediation, and long-life concrete applications. Sustainability is a celebrated topic nowadays, and as nanomaterials-based self-healing materials become incorporated into concrete products in rising amounts, it may help to develop an understanding of their interaction with the environment. Self-healing and nanotechnology have the potential to dramatically change the strength, sustainability, and whole properties of concrete. Hamers [24] reported the broad range of complex nanomaterials required to understand the molecular-level design rules. Eventually, it is challenging to exploit the power of chemistry to guarantee that nanosystems-incorporated technologies can make better environmentally friendly products.

10.3 MERITS AND DEMERITS OF NANOMATERIALS FOR SELF-HEALING CONCRETE

In general, sustainable energy and the environment are two of the central priorities for researchers, triggering a massive capital investment on research to define new trajectories in construction materials sustainability and consequent pollution abatement. Attempts by researchers to solve self-healing problems by using nanomaterials as the healing agent contributed to achieving many advantages. Moreover, deployment of nanomaterials in self-healing concrete is an emergent concept. High performance of nanomaterials positively affects the enhancement and development the self-healing concrete. The applied nanotechnology led to development of a molecular model for the hydration products (C-S-H gels) of OPC [25], as shown in Figure 10.1.

FIGURE 10.1   The C-S-H clusters: (a) TEM image and (b) molecular model generated [25].

As well as the enhancement of strength and sustainability of self-healing concrete using nanomaterials, the controlled self-healing improves and the price of the material is considerably less than that of epoxy-based materials.

The cost of self-healing concrete compared to conventional concrete still high, even when using nanomaterials. Thus, self-healing concrete is a probable product for several civil engineering structures where the concrete cost is much higher due to better quality; for instance, in tunnel linings and marine structures wherein security is a major issue, or in structures with limited accessibility for repairs and maintenance. In such special circumstances, even if costs for the concrete with self-healing agents incorporated is greater, it should not be too burdensome when looking at safety and future benefits.

10.4   ECONOMY OF NANOMATERIALS-BASED SELF-HEALING CONCRETES

Generally, concrete is the construction material most commonly used worldwide. It has been documented that more than 2.6 billion tons of OPC was manufactured in the year 2007 [26] worldwide, amounting for more than 17 billion tons. OPC was used in varieties of products such as building basements, walls, footpaths, lamp posts, bridges, dams, tall towers, skyscrapers, etc., to cite a few. Usually, concrete goods are proposed to achieve long lifetimes and as tolerant against local aggressive atmospheric conditions. Eventually, these concrete structures are mostly demolished and recycled upon reaching the final stage of their service life. Besides, in the building sector as such, it is easy to apply process innovations instead of modernizations of disruptive goods. Manufacturing combines the products from varieties of supplies in a broad array of trading into a solitary completed structure. An alteration in the built structure can be evaluated by the construction firm itself. Moreover, a considerably novel product produced by the supplier needs understanding and approval by the architects, engineers, and the client before being applied by the skilled and trained on-site workers.

Several factors must be accounted for when developing concretes incorporating nanotechnology. First, concrete and related products must be manufactured at large scale. Even if the cost of expensive concrete structures becomes lower, they must be capable of handling massive material in a safer and environmentally friendly way. Second, innovations are required to be methodically developed with field testing to achieve understanding and assurance in the construction sector. Finally, structures of concrete are hard to destroy, needing explosives or high energy for breaking up. Thus, nanotechnology-based concrete production should be compatible with these conventional practices. With these constrictions, the early nanomaterials implementations in self-healing purposes must render notable benefits in terms of extra functions with comparatively low quantities of nanomaterials. This low amount can be offered via standard construction practices and must not influence the materials' performance. Innovative products (smart self-healing concrete) must be able to advance the delivery of the traditional materials including the control of released admixtures in order to likely penetrate the marketplace.

10.5 ENVIRONMENTAL SUITABILITY AND SAFETY FEATURES OF NANOMATERIALS-BASED CONCRETES

Commonly, gaseous CO_2 is released from OPC concrete during the cement clinker's de-carbonation of lime and calcination reactions. Using nanomaterials-based self-healing technology, emission of CO_2 can remarkably be reduced. Currently, the world's development exclusive of concrete is beyond imagination [27]. History tells us that without concrete, the wonderful structures like the Sydney Opera House, the Chrysler Building, and the Taj Mahal could not have existed. Furthermore, skyscrapers in metropolises all over the world would not have reached to such striking heights without the use of celebrated concretes. In every aspect, durability of concrete played a remarkable role to erect those historic buildings centuries ago without modern technology and qualified engineers. Definitely, concrete in its own right is an integrated part of everyday life. Briefly, the manufacturing of smart and new concrete demands much money, whereas billions of tons of raw materials are wasted annually because of inefficient concrete production processes. Moreover, the production of OPC (primary concrete binder) adds up to more than 5% of the greenhouse gases released annually worldwide. It enforces the threat to our environment whereby world development is striving for sustainable and green building deployment [28].

Therefore, the future aim is targeted to build cleaner, safer, more efficient, more reliable, and stronger smart materials as alternative to concrete than the conventional OPC-based concretes. In this spirit, the notion of nanomaterials-based smart concrete and self-healing technology has been coined.

10.6 SUMMARY

The study of self-healing capability of cementitious materials has been of growing interest among the research community during the last two decades. Based on the sustainability performance of the experimental study, the following conclusions can be drawn:

i. Self-healing or smart concretes are characterized by many significant traits such as less pollution and elevated durability performance in harsh environments, as well as being cheap and eco-friendly. These properties make them effective sustainable materials in the construction industry.

ii. Use of nanomaterials in concrete is advantageous in terms of improved engineering properties of cementitious materials, especially for the generation of self-healing and sustainable concretes.

iii. Continued development and seamless integration of "self-healing" materials into modern-day applications are set to open up a new era of enhanced consumer experience, infrastructure maintenance, environmental management, and countless unprecedented and unique applications resembling those in science fiction.

REFERENCES

1. de Coninck, H.C., IPCC SR15: Summary for policymakers. In *IPCC special report global warming of 1.5 °C.* 2018: Intergovernmental Panel on Climate Change, HW Den Haag, Netherlands. p. 1–2.
2. Cobacho, S.P., et al., *Impacts of shellfish reef management on the provision of ecosystem services resulting from climate change in the Dutch Wadden Sea.* Marine Policy, 2020.**119**: p. 104058.
3. Shah, K.W., and G.F. Huseien, *Biomimetic self-healing cementitious construction materials for smart buildings.* Biomimetics, 2020.**5**(4): p. 47.
4. Huseien, G.F., et al., *Geopolymer mortars as sustainable repair material: A comprehensive review.* Renewable and Sustainable Energy Reviews, 2017.**80**: p. 54–74.
5. Goldberg, M.H., A. Gustafson, and S. van der Linden, *Leveraging social science to generate lasting engagement with climate change solutions.* One Earth, 2020.**3**(3): p. 314–324.
6. Taillardat, P., et al., *Climate change mitigation potential of wetlands and the cost-effectiveness of their restoration.* Interface Focus, 2020.**10**(5): p. 20190129.
7. Masson-Delmotte, V., et al., *Global warming of 1.5 C.* An IPCC Special Report on the Impacts of Global Warming of, 2018.**1**: p. 1–9.
8. Samadi, M., et al., *Waste ceramic as low cost and eco-friendly materials in the production of sustainable mortars.* Journal of Cleaner Production, 2020: p. 121825.
9. Mohammadhosseini, H., et al., *Enhancement of strength and transport properties of a novel preplaced aggregate fiber reinforced concrete by adding waste polypropylene carpet fibers.* Journal of Building Engineering, 2020.**27**: p. 101003.

10. Kubba, Z., et al., *Impact of curing temperatures and alkaline activators on compressive strength and porosity of ternary blended geopolymer mortars.* Case Studies in Construction Materials, 2018.**9**: p. e00205.

11. Huseien, G.F., et al., *Influence of different curing temperatures and alkali activators on properties of GBFS geopolymer mortars containing fly ash and palm-oil fuel ash.* Construction and Building Materials, 2016.**125**: p. 1229–1240.

12. Huseien, G.F., K.W. Shah, and A.R.M. Sam, *Sustainability of nanomaterials based self-healing concrete: An all-inclusive insight.* Journal of Building Engineering, 2019.**23**: p. 155–171.

13. Huseien, G.F., et al., *Effect of metakaolin replaced granulated blast furnace slag on fresh and early strength properties of geopolymer mortar.* Ain Shams Engineering Journal, 2018.**9**(4): p. 1557–1566.

14. Hu, S., et al., *Bonding and abrasion resistance of geopolymeric repair material made with steel slag.* Cement and Concrete Composites, 2008.**30**(3): p. 239–244.

15. Herbert, E.N., and V.C. Li, *Self-healing of microcracks in engineered cementitious composites (ECC) under a natural environment.* Materials, 2013.**6**(7): p. 2831–2845.

16. Rozière, E., et al., *Influence of paste volume on shrinkage cracking and fracture properties of self-compacting concrete.* Cement and Concrete Composites, 2007.**29**(8): p. 626–636.

17. Huseien, G.F., et al., *Synthesis and characterization of self-healing mortar with modified strength.* Jurnal Teknologi, 2015.**76**(1).

18. Wang, K., et al., *Permeability study of cracked concrete.* Cement and Concrete Research, 1997.**27**(3): p. 381–393.

19. Huang, H., et al., Application of sodium silicate solution as self-healing agent in cementitious materials. In *International RILEM conference on advances in construction materials through science and engineering.* 2011: RILEM Publications SARL, Hong Kong, China.

20. Wang, J., et al., *Use of silica gel or polyurethane immobilized bacteria for self-healing concrete.* Construction and Building Materials, 2012.**26**(1): p. 532–540.

21. Van Tittelboom, K., et al., *Influence of mix composition on the extent of autogenous crack healing by continued hydration or calcium carbonate formation.* Construction and Building Materials, 2012.**37**: p. 349–359.

22. Van Tittelboom, K., et al., *Self-healing efficiency of cementitious materials containing tubular capsules filled with healing agent.* Cement and Concrete Composites, 2011.**33**(4): p. 497–505.

23. Yang, Y., et al., *Autogenous healing of engineered cementitious composites under wet—dry cycles.* Cement and Concrete Research, 2009.**39**(5): p. 382–390.

24. Hamers, R.J., *Nanomaterials and global sustainability.* Accounts of Chemical Research, 2017.**50**(3): p. 633–637.

25. Pacheco-Torgal, F., and S. Jalali, *Nanotechnology: Advantages and drawbacks in the field of construction and building materials.* Construction and Building Materials, 2011.**25**(2): p. 582–590.

26. Raki, L., et al., *Cement and concrete nanoscience and nanotechnology.* Materials, 2010.**3**(2): p. 918–942.

27. Onaizi, A.M., et al., *Effect of nanomaterials inclusion on sustainability of cement-based concretes: A comprehensive review.* Construction and Building Materials, 2021.**306**: p. 124850.

28. Samadi, M., et al., *Influence of glass silica waste nano powder on the mechanical and microstructure properties of alkali-activated mortars.* Nanomaterials, 2020.**10**(2): p. 324.

Index

A

admixtures, 24, 35, 44, 101–115, 120, 134
agents, 3, 5, 13–16, 24, 31–33, 46, 94, 107
aggressive environment, 1, 4, 6
alkaline environment, 88–90, 116, 125
autonomous, 2, 6, 33, 41–49, 87–88, 113

B

Bacillus, 34, 89, 91–93, 123–124, 127
bacteria, 5, 13, 27, 34, 42–48, 56, 73, 87–94
biotechnology, 41, 87, 99
building problems, 11, 23, 87

C

carbonate, 5, 13, 24, 27, 42–48, 88, 116
carbonation, 11, 13, 136
carbon nanotube, 35, 100, 102
cement, 1–6, 11, 21, 23, 56, 87, 133
cementitious, 4–6, 12, 24, 88, 134
climate change, 1, 4, 6, 11, 137
coating, 14, 18, 30, 31, 106, 134
concrete, 1–5, 12, 21, 43, 88, 135
corrosion, 3, 12, 30, 33, 56, 99–100, 106
cracks, 2–6, 15, 21, 30, 42, 47, 68
crystalline, 21, 57, 105, 113–115, 120

D

damage, 2, 12, 14, 21–25, 30, 35
deterioration, 3, 6, 11, 14, 25, 30, 55, 91
development, 2, 3, 6, 12, 27, 30, 42, 56, 64
dosage, 5, 46, 101, 104, 117, 127
durability, 3, 11, 14, 18, 22, 30, 34, 44, 48

E

eco-friendly, 12, 34, 107, 137
economic, 5, 11, 13, 123, 133
economically, 3, 106
economic benefits, 123
efficiency, 46, 69, 74
encapsulate, 6, 15, 21, 44, 47, 90, 126
environment, 1–6, 13, 15, 23, 30, 34, 71, 84
environmental safety, 5, 13
epoxy, 26, 30, 31, 45
epoxy resins, 16, 26, 56

G

gels, 56, 57, 66, 73, 102, 134
green environment, 5
greenhouse, 1, 2, 133, 136

H

harsh environment, 17, 34, 107, 128, 137
hollow fibers, 15, 17, 24–26, 35, 42, 102
hydration, 5, 13, 21, 24, 42, 48, 55

I

immersion, 21, 62, 113–118, 121, 128
improvement, 3, 6, 24, 26, 33, 66, 87, 91
infrastructure, 3, 6, 12, 23, 30, 41, 48, 137
injection, 26, 43, 44, 102

L

lactate, 34, 90, 91, 123–125, 128
life cycle assessment, 6

M

macroscale, 27, 46
macrostructure, 47, 49
maintenance, 5, 14, 22, 41, 55, 71, 123, 137
microbial, 44, 89, 91
microbiologically, 88
microcapsule, 5, 15, 28, 31, 46, 90, 117
morphology, 68, 72, 74, 98

N

nanoalumina, 98, 99, 100, 102
nanocomposite, 43, 98, 105
nanomaterials, 1, 33, 43, 97, 100, 107, 134
nanosilica, 98, 100, 101
nanostructure, 2, 12, 46, 47, 50
nanotechnology, 12, 35, 43, 98, 136

O

organism, 25, 47, 88

P

permeability, 21, 44, 48, 55, 87, 91, 93, 104
phenomena, 11, 27
polymer, 15–18, 24, 31, 43
polymerization, 24, 28, 31, 64, 67, 84
porosity, 48, 56, 66, 68, 84
pre-cracks, 114
pre-loading, 42, 68, 69, 71
pure, 56, 103, 127

R

reaction, 3, 13, 21, 27, 55, 56, 66, 68, 73
resistance, 1, 16, 22, 29, 33, 56, 88

S

self-healing, 1–6, 12–16, 31–36, 44,
        57, 136
self-repair, 3–6, 14, 26, 29, 30
silica, 24, 47, 57, 59, 67, 89, 93, 99
smart concrete, 1, 5, 11, 99, 103, 137
strength, 3, 18, 22, 30, 35, 43, 48, 56
sustainability, 1, 5, 13, 107, 129,
        131–137

W

water absorption, 44, 61, 68, 83, 87, 92

Printed in the United States
by Baker & Taylor Publisher Services